Experiments in Introductory Chemistry

Experiments in Introductory Chemistry

Karen Pressprich
Indiana University

R. A. D. Wentworth
Indiana University

Introductory Chemistry

Darrell D. Ebbing
Wayne State University

R. A. D. Wentworth
Indiana University

with James P. Birk
Arizona State University

BROOKS/COLE
CENGAGE Learning

Australia • Brazil • Japan • Korea • Mexico • Singapore • Spain • United Kingdom • United States

For product information and technology assistance, contact us at
Cengage Learning Customer & Sales Support, 1-800-354-9706.

For permission to use material from this text or product, submit all requests online at
www.cengage.com/permissions.
Further permissions questions can be emailed to
permissionrequest@cengage.com.

ISBN-13: 978-0-395-46626-1
ISBN-10: 0-395-46626-1

Brooks/Cole
20 Davis Drive
Belmont, CA 94002-3098
USA

Cengage Learning is a leading provider of customized learning solutions with office locations around the globe, including Singapore, the United Kingdom, Australia, Mexico, Brazil, and Japan. Locate your local office at: **www.cengage.com/global**.

Cengage Learning products are represented in Canada by Nelson Education, Ltd.

To learn more about Brooks/Cole, visit
www.cengage.com/brookscole.

Purchase any of our products at your local college store or at our preferred online store **www.cengagebrain.com**.

Printed in the United States of America
5 6 7 8 9 10 11 12 13 12 11 10

Contents

Preface

Experiments in Introductory Chemistry is written to accompany *Introductory Chemistry* by Darrell D. Ebbing, R. A. D. Wentworth, and James P. Birk. Although this laboratory manual can be used with other textbooks, it uniquely complements the organization and coverage of the Ebbing/Wentworth/Birk text. As a result, an immediate advantage is realized: The lecture and laboratory parts of a course can be integrated with relative ease.

Organization and Content

The twenty-two experiments in this manual are arranged to parallel the material found in the twenty chapters of *Introductory Chemistry.* At least one experiment is provided for each chapter of the text. For two chapters, two experiments have been written, allowing better coverage of important concepts if time permits, or a choice of topics if it does not. The experiments range from qualitative inspections of various reactions to quantitative interpretations of stoichiometry, analysis, and many chemical phenomena.

Each experiment is numbered according to the chapter with which it can be used. For example, Experiment 1 can be used with Chapter 1, Experiments 14A and 14B can be used with Chapter 14.

A student will not need to refer to the textbook to understand the laboratory manual. All required explanations and data are given in this manual.

Each experiment is carefully organized, with clearly stated purpose, in-depth discussion of the experiment in the introductory section, and step-by-step procedures. All experiments include a prelaboratory assignment that should be completed by students before coming to the lab. Separate sections are provided for noting the results obtained during the experiment and for questions to be answered after completion of the lab. All three sections—prelaboratory assignments, results, and questions—are on perforated pages, which can be easily detached and handed in.

Safety issues receive special attention throughout the manual. General safety rules are given in the Introduction. As is appropriate, students are reminded of these rules in each experiment as well as given special precautions that have been formulated for the particular experiment. These are highlighted by the heading "SAFETY PRECAUTIONS." All of these precautions have been thoroughly reviewed.

Each experiment has been tested by literally thousands of students for years. All of the experiments can be completed in about two hours.

Acknowledgments

Many wonderful people have helped us with this laboratory manual. We wish to acknowledge the freshmen and teaching assistants at Indiana University; and the manual's reviewers Richard Reiter, Illinois State University; Chhiu Tsu Lin, Northern Illinois University; Floyd W. Kelly, Casper College. K. A. P. also acknowledges Steve, Rachael, and Jessica for their incredible patience and support during the time that the manual was written.

Introduction

The study of chemistry is a fundamental part of any science curriculum. Since chemistry has developed largely through experiments, the study of chemistry is augmented by laboratory experiences that demonstrate, clarify, and develop the principles of chemistry discussed in the classroom. This laboratory manual, along with guidance by your laboratory instructor, will help you with the first steps toward the realization of these objectives. However, you must shoulder a significant portion of the burden.

Once you learn to follow certain codes of conduct pertaining to safety and housekeeping in the laboratory, you will be able to appreciate your laboratory experiences to a greater degree. You will be able to complete the experiments described in this manual in a relatively safe environment. Good housekeeping practices will also help to prevent contamination of the laboratory chemicals. Good safety and housekeeping pratices, along with correct use of glassware and other equipment, will ensure success and safety with these experiments.

Safety Rules

Work in a laboratory should be a safe experience. It will be safe, however, only if certain safety precautions are followed without exception. Before you read the following 18 safety rules, it is worthwhile to consider their common origin: each rule is the direct result of several accidents. Only with hindsight is the question asked, "What regulation would have prevented these accidents?" Once each rule has been formulated, it seems so reasonable and so much in accord with common sense that we can only wonder why it was not prescribed before the accidents occurred.

The safety rules that must be followed in the chemistry laboratory are listed below.

1. *Locate the safety equipment.* Find the eye wash fountains, safety showers, fire extinguishers, fire blankets, first aid kit, emergency telephone, and all exits that are to be used in an emergency. Your laboratory instructor will describe the use of the safety equipment.
2. *Protect your eyes.* Wear safety glasses or goggles at all times. Goggles provide maximum safety. Prescription eyeglasses, if you need them, must be worn *under* safety glasses or goggles. Do not wear contact lenses because various fumes may accumulate under the lenses and injure your eyes.
3. *Tie long hair back.* This precaution will keep your hair out of burner flames and harmful liquids.
4. *Wear shoes that completely cover your feet.* Since broken glass and chemical spills occur all too often, your feet require more protection than is provided by open-toed shoes or sandals.
5. *Wear clothes that cover your torso and your legs to the knees.* Clothing will protect your body against chemical spills.
6. *Do not eat or drink in the laboratory.*
7. *Do not taste any chemical.*
8. *Do not directly inhale any chemical.* Use your hand to waft the odor to your nose.
9. *Do not pipet solutions by mouth.* Use a rubber suction bulb or special pipet filler.
10. *Do not place flammable liquids near an open flame.*
11. *Do not engage in horseplay in the laboratory.*
12. *Do not do unauthorized experiments.*
13. *Do not work in the laboratory in the absence of your laboratory instructor or his or her authorized representative.*
14. *Use a hood when required.* This manual will indicate when a hood (or other similar device) is needed.
15. *Handle glass tubing with care.* Glass tubing is easily broken. When tubing (including thermometers) is to be inserted through a rubber stopper, the tubing must be lubricated with water or glycerol. Hold the tubing with a

cloth or a paper towel near the end that will be inserted, and use a twisting motion during insertion.

16. *Be aware of your neighbors. Are they obeying the safety rules?* A neighbor's accident may not hurt him or her but may injure you.
17. *Report any accident or spill to your laboratory instructor.*
18. *Do not violate any other safety rule given in this manual or issued by your laboratory instructor.*

Housekeeping Rules

Good housekeeping in the laboratory will provide a safe work site in which you may be assured that chemicals are not contaminated. Please observe the following rules.

1. *Clean up broken glass immediately.* Special containers may be provided for disposal.
2. *Clean up solid and liquid spills immediately.* Report any spill to your laboratory instructor.
3. *Do not pour any chemical into a sink without authorization.* Often, disposal bottles will be provided.
4. *Take containers to the stock of chemicals.* Do not bring stock chemicals to your laboratory bench.
5. *Read the label on a bottle carefully.* Is it the correct chemical?
6. *Do not insert a pipet or medicine dropper directly into a stock bottle.* Avoid contamination by pouring the liquid into a test tube, flask, or beaker before taking a sample.
7. *Use special care with stoppers or tops of stock bottles.* Do not allow them to be contaminated.
8. *Take no more of a chemical than the experiment requires.*
9. *Never return an unused chemical to a stock bottle.* You must assume that the chemical is contaminated. It must be discarded.
10. *Set up your glassware and apparatus at least 6 inches from the front edge of your laboratory bench.*
11. *Follow any other housekeeping rules given by your laboratory instructor.*

Common Glassware and Other Equipment

The glassware and other equipment found in most general chemistry laboratories is shown in Figure I.1. Refer to this figure whenever an experiment calls for a piece of unfamiliar equipment. We will now discuss the use of certain pieces of this equipment.

Figure I.1
Common glassware and equipment.

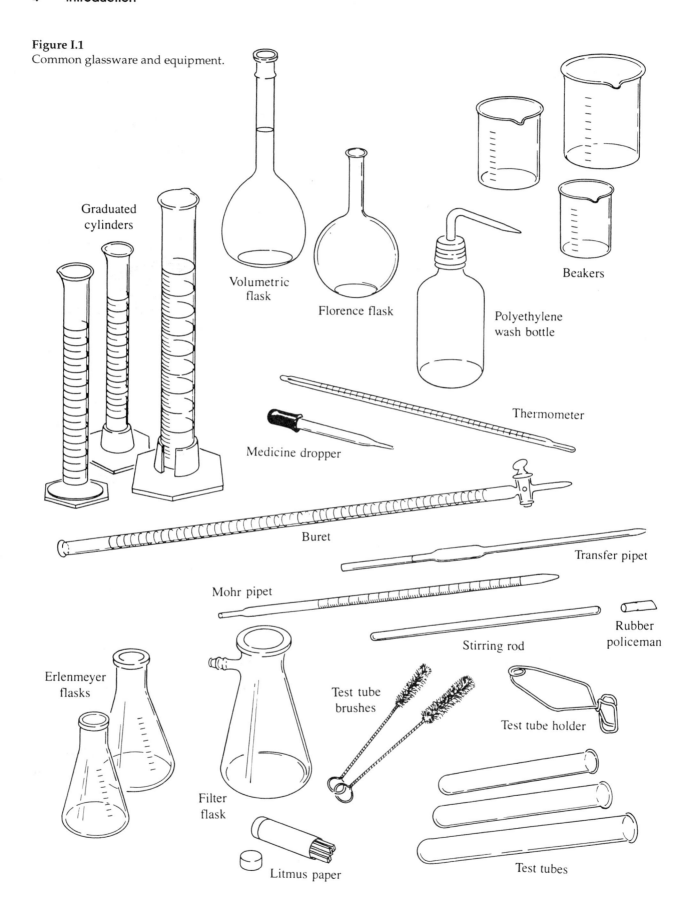

Graduated
cylinders

Volumetric
flask

Florence flask

Polyethylene
wash bottle

Beakers

Medicine dropper

Thermometer

Buret

Transfer pipet

Mohr pipet

Stirring rod

Rubber
policeman

Erlenmeyer
flasks

Filter
flask

Test tube
brushes

Test tube holder

Litmus paper

Test tubes

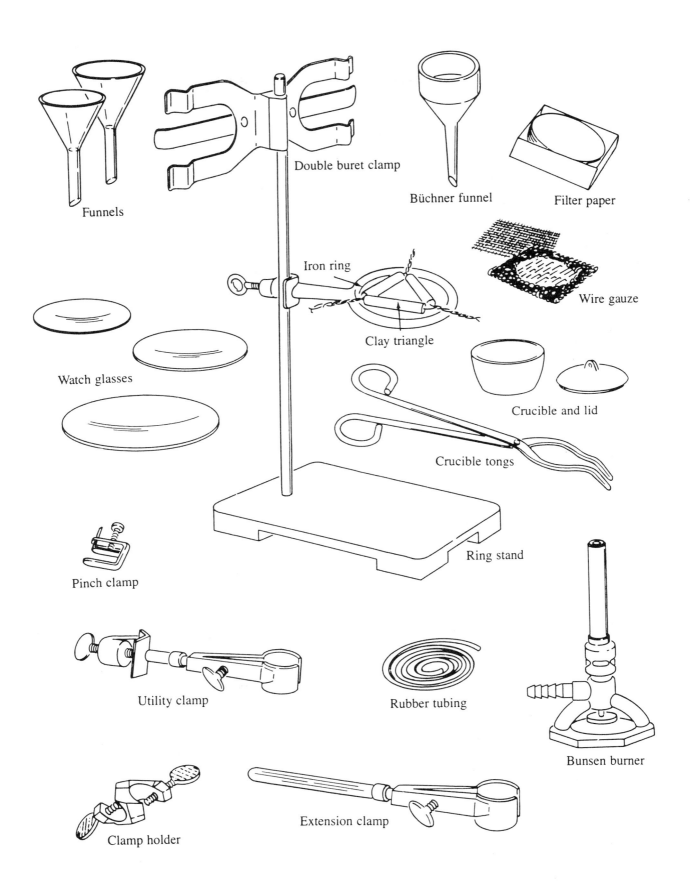

Funnels

Double buret clamp

Büchner funnel

Filter paper

Iron ring

Wire gauze

Clay triangle

Watch glasses

Crucible and lid

Crucible tongs

Ring stand

Pinch clamp

Utility clamp

Rubber tubing

Bunsen burner

Clamp holder

Extension clamp

Figure I.2
The proper method of reading a meniscus so that parallax error is avoided.

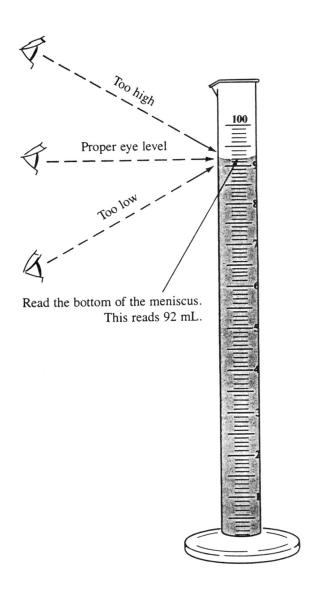

Too high

Proper eye level

Too low

Read the bottom of the meniscus.
This reads 92 mL.

Volumetric Glassware

Graduated cylinders, transfer pipets, Mohr pipets, and burets are used to make volumetric measurements. This glassware must be scrupulously clean. Liquids must drain without leaving drops adhering to the inner walls of the glassware.

Graduated Cylinders. A graduated cylinder is used to measure an approximate volume of liquid. When water or an aqueous solution (a solution containing water as the solvent) is added to the graduated cylinder, the upper surface of the liquid will be concave. This concave surface is called a *meniscus.* The bottom of the meniscus is used for all measurements. To avoid error (called *parallax error*), your eye should always be level with the meniscus when you are measuring the volume. Examine Figure I.2 with care.

Some graduated cylinders are calibrated *to contain* (*TC*) the volume that is measured. You will usually find TC etched in the glass of this type of graduated cylinder. The delivered volume will always be slightly less than the measured volume because of the residual liquid that coats the inner walls and remains after the liquid is poured out. Other graduated cylinders are calibrated *to deliver* (*TD*) a measured volume. The actual volume of a liquid in this type of cylinder is always slightly greater than the measured volume to compensate for the residual liquid.

Graduated cylinders come in many sizes. The 10-mL, 25-mL, 50-mL, and 100-mL graduated cylinders are often found in general chemistry laboratories. You will be able to measure any volume up to the maximum volume of the graduated cylinder.

Pipets. Transfer and Mohr pipets are required for some of the experiments in this laboratory manual. A transfer pipet is calibrated to deliver (TD) only one volume, whereas a Mohr pipet is graduated so that it can deliver any volume (usually to the nearest tenth of a milliliter) up to its maximum volume. Transfer pipets come in many sizes; 5-mL, 10-mL, and 20-mL pipets are usually found in general chemistry laboratories. Mohr pipets are commonly restricted to 5-mL and 10-mL volumes.

The correct use of a pipet requires considerable manipulatory skill. This is not an innate skill, but one that comes only with practice. Step-by-step procedures for correct usuage with a rubber suction bulb are shown in Figure I.3. If you are to use other suction devices, your laboratory instructor will discuss and demonstrate them. Although the figure shows a transfer pipet, the instructions also apply to a Mohr pipet. Remember that you must not use your mouth for suction even if you are filling the pipet with water!

Burets. The principal use of the buret is for titration, as shown in Figure I.4. Precise titrations require burets that drain freely, are very clean, and do not leak around the stopcock. The following three steps will help you to have a buret that operates as it should.

1. The capillary tip of the buret should be clean and free of foreign objects. A thin wire can sometimes be used successfully to dislodge grease or dirt that impairs or prevents draining.
2. If water droplets are left on the inner walls of the buret after draining, the buret needs a thorough cleaning. It should be cleaned with hot water, detergent, and a buret brush; then it should be rinsed with tap water. Finally, it should be rinsed with distilled water.
3. Some maintenance is required if the stopcock leaks while the buret is draining or if drops form on the capillary tip when the stopcock is turned off. Glass stopcocks must be lubricated to prevent both kinds of leaking. Lubrication will also allow the stopcock to turn easily. Your laboratory

Figure I.3
The correct procedure for using a pipet with a rubber suction bulb.

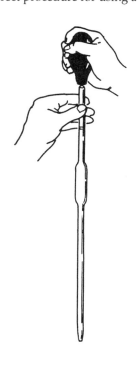

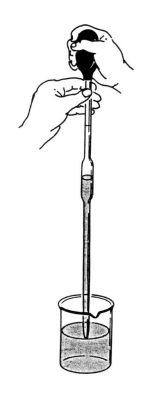

(a) Place the rubber suction bulb at the mouth of the pipet, using your more dexterous hand to hold the bulb. Do not insert the pipet into the bulb. Squeeze the air from the bulb.

(b) Insert the pipet into the liquid. Allow the bulb to expand *slowly* to draw liquid into the pipet. Do not allow the liquid to enter the bulb, where it will be contaminated.

(c) When the liquid is about 1 cm above the etched line on the pipet, remove the bulb and place the tip of your index finger from your less dexterous hand over the mouth of the pipet. Remove the pipet from the liquid, and dry its exterior with tissue paper.

instructor will show you how to grease the stopcock without allowing the grease to enter the tip of the buret. Teflon stopcocks do not require lubrication. Leaking of Teflon stopcocks can usually be prevented by tightening the tension nut, which seats the stopcock more firmly.

Laboratory Burners

Your laboratory will probably be equipped with Bunsen burners like the one shown in Figure I.5. If you have another type of burner, your laboratory instructor will describe its use.

Operating Your Bunsen Burner. Your Bunsen burner relies on the combustion of natural gas or bottled gas. To achieve the best flame, the gas inlet valve and

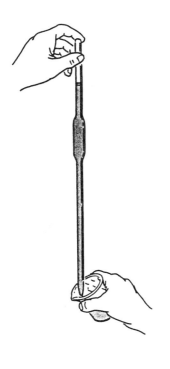

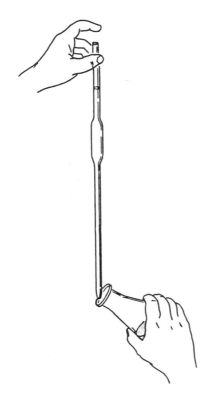

(d) Drain the excess into a waste container until the bottom of the meniscus coincides with the etched line. Touch off any adhering last drop.

(e) Allow the pipet to drain into the container to be used in the experiment. Again, touch off any last drop.

(f) Do not blow the remaining liquid from the pipet. The pipet was calibrated to deliver the correct volume with this liquid remaining in it.

the air vents must be properly adjusted. If you adopt the following procedure, you should be able to obtain the best flame for your burner without difficulty.

Your burner should be equipped with a short piece of rubber tubing. One end of the tubing should be attached to the sidearm of the burner, and the other end should be connected to an outlet from a source of gas. Open the gas outlet completely. Strike a match, and open the gas inlet valve on the burner about halfway. Light the burner by bringing the match from the side to the top of the burner. With any other approach, the match may be extinguished by the flow of gas.

Close the air vents and adjust the gas inlet valve until the flame is about 4 inches high. The flame should be luminous and yellow. Open the air vents until the flame becomes two concentric cones. The outer cone will be only faintly

Figure I.4
Technique for titration by a right-handed student.

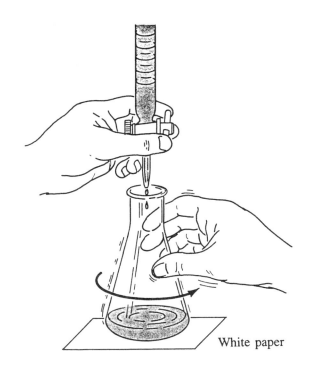

White paper

Figure I.5
A Bunsen burner.

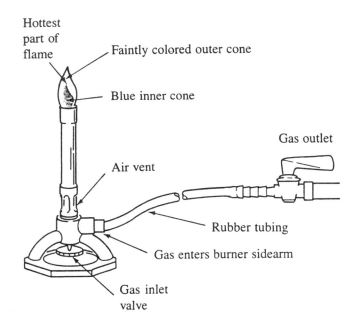

Hottest part of flame

Faintly colored outer cone

Blue inner cone

Gas outlet

Air vent

Rubber tubing

Gas enters burner sidearm

Gas inlet valve

colored, but the inner cone will be blue. The hottest part of the flame is at the tip of the blue cone.

Filtration. Filtration through special paper (called filter paper) is the simplest method of separating a solid from a liquid. Filter paper is available with a variety of porosities. A finely porous paper should be used for solids made up of very small particles, but filtration will be slow. A coarser, more porous paper can be used with solids whose particles are larger. Filtration will then be more rapid.

Gravity Filtration. This technique requires a conical filter funnel with a hollow stem, and a glass stirring rod. The stirring rod may be equipped with a rubber policeman. This device, which slips over the end of the stirring rod, is used to remove precipitates from the walls of a beaker or flask.

For this method of filtration, the filter paper is folded in the manner shown in Figure I.6. The small tear in the corner of the outside fold permits a closer seal with the funnel during filtration. After you make this paper cone, place it in the filter funnel and wet the paper thoroughly with distilled water. Pour off the excess water and place the funnel in its support. The support can be an iron ring, a clay triangle on an iron ring, a wooden board with a circular hole, or the mouth of an Erlenmeyer flask. If you are filtering into a beaker, you can minimize splashing by putting the stem of the filter funnel against the inner wall of the beaker.

Figure I.6
The correct method of folding filter paper for gravity filtration.

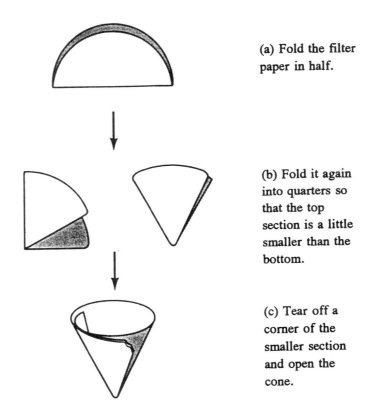

(a) Fold the filter paper in half.

(b) Fold it again into quarters so that the top section is a little smaller than the bottom.

(c) Tear off a corner of the smaller section and open the cone.

The next step is shown in Figure I.7. The mixture to be filtered should be poured (*decanted*) along the stirring rod to direct the flow of the liquid into the filter paper. Never fill the filter paper to more than two-thirds of its volume. Most of the solid should be transferred to the filter during this stage. If you have a rubber policeman, use it to remove any solid that adheres stubbornly to the walls of the container.

When the liquid has been transferred to the filter, use a stream of distilled water from a plastic wash bottle to rinse the remainder of the solid into the paper. This technique is shown in Figure I.8. When all the solid has been transferred, rinse the stirring rod in such a way that the distilled water is also directed onto

Figure I.7
The correct method
of decanting into
the filter paper.

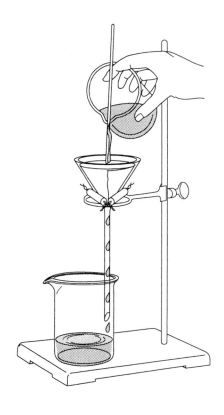

Figure I.8
Use a stream
of distilled water
from a plastic wash bottle
to transfer all
of the solid
into the filter paper.

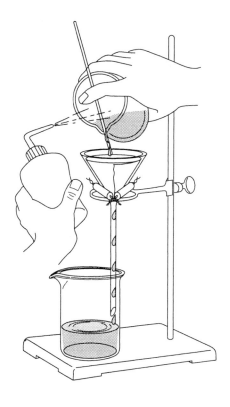

the paper. Finally, wash the solid with two small portions of distilled water. A quantitative transfer of the solid has now occurred.

Suction Filtration.   This type of filtration is much faster than gravity filtration, but quantitative recovery of a solid is rarely achieved. Suction filtration requires a Büchner funnel, a suction flask, a rubber stopper or rubber ring to hold the funnel tightly in the flask, a glass stirring rod, heavy rubber tubing, and a water aspirator. In addition, some type of safety trap is advised to prevent water from the aspirator from backing up into the suction flask.

The Büchner funnel and suction flask should be set up as shown in Figure I.9. Clamp the flask to a ring stand. Place a piece of filter paper of the correct size in the funnel. Turn on the water aspirator to start the suction. Wet the filter paper with distilled water and allow the water to drain into the suction flask; then empty the flask after the water has drained. Place the funnel on the flask once again.

Transfer as much of the solid and liquid to be filtered into the funnel as you can. Use the stirring rod to help the transfer. Repeat this operation until all of the liquid has passed through the funnel. Transfer the filtered liquid back to its original container, and repeat the operation. Eventually, all of the solid will be transferred into the funnel. Wash the solid with distilled water or any other liquid that is required, and suction dry.

Figure I.9
The correct setup for suction filtration.

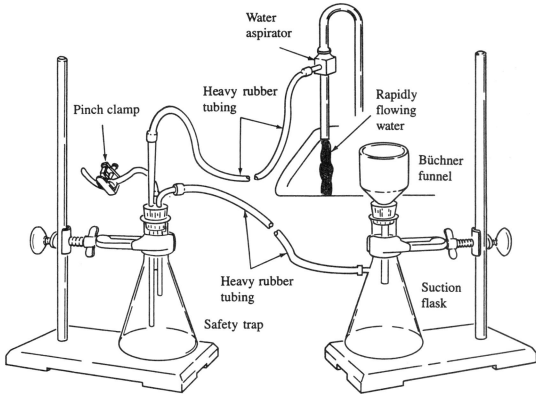

1
Identification of an Unknown Compound

Objectives

- Learn to make accurate and complete observations.
- Identify an unknown compound by comparing its reactions with those of some known compounds.

Introduction

Chemistry is a science that is built on the interrelationship of experiment and theory. Accurate and complete observations are required during experiments to provide the maximum amount of useful information. Good observations require practice and attention to detail.

You will obtain solid samples of sodium chloride (NaCl), sodium carbonate (Na_2CO_3), sodium hydrogen phosphate (Na_2HPO_4), and sodium sulfate (Na_2SO_4), as well as solutions of these compounds in water. You will test these compounds for reactions with solutions of nitric acid (HNO_3), barium nitrate [$Ba(NO_3)_2$], and silver nitrate ($AgNO_3$). You must observe each characteristic reaction carefully. Simultaneously, you will also test an unknown sample that will be identical to one of the solid samples. You will be able to determine its identity by matching its characteristic reactions with those of the known compounds.

A **chemical reaction** occurs when substances are converted into new products having properties and compositions that are entirely different from those of the starting materials. During this experiment you will find that there are signals that indicate when chemical reactions have taken place. These signals include:

1. The formation or disappearance of a **precipitate,** which is a solid product. A precipitate is often made up of very small solid particles evenly distributed throughout the solution, giving it a cloudy appearance. An example of a precipitate is shown in Figure 1.1.
2. The formation of a gaseous product. When gas is produced, bubbles often form in solution, a process known as **effervescence.** A change in the smell of the solution also indicates that a gas has been produced.

   CAUTION: Always be very careful with solutions that are bubbling. They may become hot, and the gas may be strong-smelling and/or toxic. Never smell chemicals directly. Use your hand to waft the odor to your nose.

Experimental

Equipment: 5 small (13 mm × 100 mm) test tubes
red and blue litmus paper
100-mL graduated cylinder
400-mL beaker
stirring rod
medicine dropper
marking pencil
spatula

Figure 1.1
The formation of a
cloudy precipitate
when one solution
is added to another.

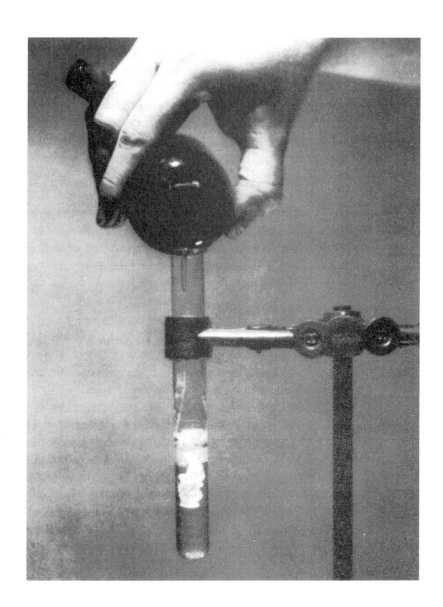

Materials: solid NaCl (sodium chloride)
0.1 *M* NaCl
solid Na₂CO₃ (sodium carbonate)
0.1 *M* Na₂CO₃
solid Na₂HPO₄ (sodium hydrogen phosphate)
0.1 *M* Na₂HPO₄
solid Na₂SO₄ (sodium sulfate)
0.1 *M* Na₂SO₄
0.1 *M* Ba(NO₃)₂ (barium nitrate)
0.1 *M* AgNO₃ (silver nitrate)
6.0 *M* HNO₃ (nitric acid)
6.0 *M* NH₃(*aq*) (aqueous ammonia, also known as ammonium
hydroxide or NH₄OH)

Safety Precautions

Nitric acid and aqueous ammonia can cause chemical burns in addition to ruining your clothing. If you spill either of these solutions on you, wash the contaminated area thoroughly with tap water and report the incident to your laboratory instructor. You may require further treatment. Aqueous ammonia has a very strong smell so you should avoid directly inhaling ammonia.

Procedure

Getting Started

1. Obtain the unknown compound from your laboratory instructor.
2. You will need 5 test tubes, a medicine dropper, and red and blue litmus paper.
3. Obtain directions from your laboratory instrutor for discarding the solutions that you will use in this experiment.

Testing for Gas Evolution

1. Identify your test tubes with a marking pencil. (Example: Cl^-, CO_3^{2-}, HPO_4^{2-}, SO_4^{2-}, and U, standing for $NaCl$, Na_2CO_3, Na_2HPO_4, Na_2SO_4, and unknown, respectively.)
2. Use a clean spatula to place pea-sized solid samples of $NaCl$, Na_2CO_3, Na_2HPO_4, Na_2SO_4, and your unknown in the test tubes.
3. Add 5 drops of the HNO_3 solution carefully to each test tube and record your observation.
4. Discard the solution in the test tubes.
5. Wash the test tubes and rinse them with distilled water.

Dissolving Your Unknown Compound

1. Wash a 100-mL graduated cylinder and a 400-mL beaker thoroughly and rinse them with distilled water.
2. Place a pea-sized portion of your unknown in the beaker. Add 200 mL of distilled water from the graduated cylinder, and swirl or stir gently until all of the solid has dissolved.
3. Set aside the remaining portion of your unknown compound in a safe place for use in the event of an unforeseen accident.

Testing with Barium Nitrate

1. Use the solutions of $NaCl$, Na_2CO_3, Na_2HPO_4, and Na_2SO_4 that you will find in the laboratory and the solution of your unknown compound that you have just prepared for the remaining tests in this experiment.
2. Place 20 drops of the $NaCl$ solution in a clean, correctly marked test tube. Add 3 drops of a solution of $NH_3(aq)$. Use a clean, dry stirring rod to stir the solution. Remove the stirring rod, and touch the adhering drop of solution to a small piece of red litmus paper. If the paper does not turn blue, add $NH_3(aq)$ dropwise to the solution in the test tube until the litmus paper turns blue.
3. Repeat step 2 with each of your solutions.

4. Add 5 drops of the Ba(NO$_3$)$_2$ solution to each test tube. Shake gently to obtain homogeneity. Do not use your finger as a stopper. Examine each test tube carefully and look for insoluble precipitates. Record your observations, noting the color of each precipitate.

5. The test in this step applies only to those test tubes that contain precipitates. Add 10 drops of the HNO$_3$ solution to each of these test tubes. Test each solution with blue litmus paper, using a clean, dry stirring rod. If the paper does not turn pink, add more drops of the HNO$_3$ solution to the solution in the test tube until the litmus paper turns pink. Examine each test tube. Which precipitates have dissolved? Record your observations.

6. Discard the solutions in the test tubes.

7. Wash the test tubes and rinse them with distilled water.

Testing with Silver Nitrate

1. Use fresh solutions for this test.

2. Add 20 drops of the NaCl solution to a clean, correctly marked test tube.

3. Repeat step 2 with each of your solutions.

4. Add 5 drops of the AgNO$_3$ solution to each test tube. Record your observations.

5. The test in this step applies only to those test tubes that contain precipitates. Add 10 drops of the HNO$_3$ solution to each of these test tubes. Test each solution with blue litmus paper, using a clean, dry stirring rod. If the paper does not turn pink, add more drops of the HNO$_3$ solution until the solution in the test tube does turn the litmus paper pink. Did the precipitates dissolve? Record your observations.

6. Discard the solutions in the test tubes. Wash and rinse the test tubes using distilled water.

Identifying the Unknown Compound

1. With one possible exception, the characteristic reactions of the unknown compound should be identical to those of one of the known samples. If your unknown is Na$_2$CO$_3$, its behavior in solution when tested with AgNO$_3$ may differ slightly from that of a known sample. Differences in the concentrations of these solutions can lead to reactions that result in slightly different colors. Keeping this possible difference in mind, you will be able to identify your unknown compound by matching its reactions with those of one of the known samples.

2. If ambiguities occur, repeat as many of the tests as you find necessary.

Identification of an Unknown Compound

NAME _____

SECTION _____

DATE _____

Prelaboratory Assignment

1. Define the following terms:
   a. precipitate

   b. effervescence

2. Match the name of each compound with its formula.

sodium hydrogen phosphate	Na_2CO_3
ammonia	HNO_3
sodium carbonate	$NaCl$
sodium sulfate	Na_2HPO_4
nitric acid	Na_2SO_4
sodium chloride	NH_3

3. List the signals that indicate a chemical reaction has occurred.

4. What special safety precautions must be observed during this experiment?

Identification of an Unknown Compound

NAME _____

SECTION _____

DATE _____

Results

	HNO$_3$	Ba(NO$_3$)$_2$	Ba(NO$_3$)$_2$ + HNO$_3$	AgNO$_3$	AgNO$_3$ + HNO$_3$
NaCl					
Na$_2$CO$_3$					
Na$_2$HPO$_4$					
Na$_2$SO$_4$					
Unknown no. _____					

My unknown is _____ .

Question

1. An unknown sample contains at least two of the four compounds that you handled in this experiment. The following observations were made:

    a. No gas evolves when the solid is treated with HNO_3.
    b. A white precipitate forms after the addition of solutions of $NH_3(aq)$ and $Ba(NO_3)_2$ to a solution of the unknown. The precipitate does not dissolve after the addition of a solution of HNO_3.
    c. A white precipitate forms after the addition of a solution of $AgNO_3$ to a solution of the unknown. The precipitate does not dissolve after the addition of a solution of HNO_3.

    Use your observations to identify the components of the unknown sample. Explain.

2

Determination of Mass, Volume, and Density

Objectives

- Develop proficiency in the use of the metric system.
- Become familiar with some of the common measuring devices in the laboratory.
- Determine the density of a solid.
- Determine the density of pure liquids and solutions.

Introduction

Density, like color, odor, solubility, melting point, and boiling point, is a physical property of a substance. Therefore, density may be used in identifying a substance. **Density** of a substance is defined as its mass per unit volume. Mathematically, density can be expressed as: $d = m/V$, where d is density, m is mass, and V is volume.

In the metric system, the unit of mass is the gram (g), the unit of volume for a solid is the cubic centimeter (cm^3), and the unit of volume for a liquid is the milliliter (mL). Therefore, the density of a solid or liquid is usually expressed in grams per cubic centimeter (g/cm^3) or grams per milliliter (g/mL), respectively.

When we say that gold (Au) is more dense than aluminum (Al), we mean that a gold cube or block is heavier (has a larger mass) than an aluminum cube or block of the same size. For example:

Volume of aluminum = 1 cm^3	Volume of gold = 1 cm^3
Weight of aluminum = 2.70 g	Weight of gold = 19.7 g
Density of aluminum = 2.70 g/cm^3	Density of gold = 19.7 g/cm^3

When two substances are mixed, the mixture may be completely uniform in appearance. In this case the liquids (or gases) are said to be **miscible.** If the two liquids (or gases) are **immiscible,** two distinct layers will form when they are poured together. The layer on the bottom always has the greater density. For example, a balloon filled with helium gas rises in air because helium gas has a lower density (0.178 g/L) than air (1.293 g/L). Vinegar and olive oil are immiscible; the vinegar, being denser, forms the bottom layer.

Density can also be used to determine the concentration of solutions in certain instances. When a solute is dissolved in a solvent, the density of the solution will be different from that of the pure solvent itself.

Determination of the density of certain physiological liquids is often an important screening tool in medical diagnosis. For example, if the density of urine differs from normal values, this may indicate a problem with the kidneys secreting substances that should not be lost from the body. The determination of density is almost always performed as part of a urinalysis. Another interesting example utilizing density is the determination of total body fat. Muscle is more dense than fat; therefore, by determining total body mass and volume, the muscle-to-fat ratio can be calculated.

In this experiment you will determine densities of various substances by measuring their volumes with a graduated cylinder and their masses with a balance. You will also determine the concentration of sodium chloride dissolved in water in a sample of unknown concentration based on its experimentally determined density.

Graphing

Much of the data you collect in this experiment will be presented in the form of a graph. Here are some pointers for making successful graphs:

1. Title the graph, and label all the axes with units (grams, cubic centimeters, etc.).
2. Spread the axes out so that the data covers the entire graph. You will need to decide on the size of divisions to mark on your graph.
3. Consider what these divisions will mean in terms of each box on the graph paper. A common student error is to make unequal divisions.
4. Plot all points carefully and neatly. Use a pencil for graphing.
5. Do not "connect-the-dots"! If your graph represents a straight line (as for the calibration curve), use a ruler and draw the best straight line through your data points. This line does not have to intersect all the data points. Also, your line does not have to go through the origin (also known as the zero–zero point). If your data are scattered, estimate where to draw your best straight line. Roughly an equal number of points should be above the line and below the line. This approximates a mathematical technique called *linear regression*, which judges where to draw the line to minimize the distance from each point to the line.
6. When you are reading data (for your unknown) from the graph, it is a good idea to use a ruler to read the data point by drawing a dotted line from the *x*-axis to the point. Do the same from the point to the *y*-axis.

Some do's and don't's in graphing follow.

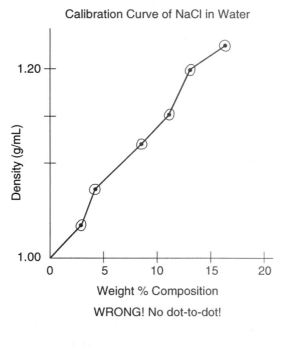

Calibration Curve of NaCl in Water

WRONG! No dot-to-dot!

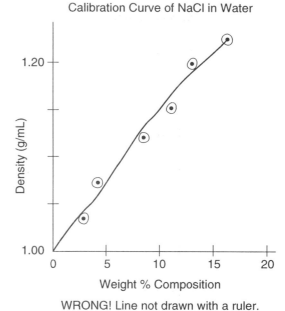

Calibration Curve of NaCl in Water

WRONG! Line not drawn with a ruler.

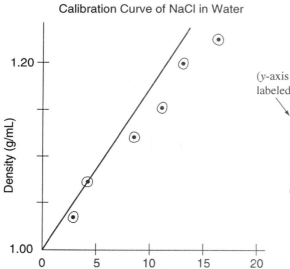

Calibration Curve of NaCl in Water

WRONG! Line does not have half the
points on each side of it.

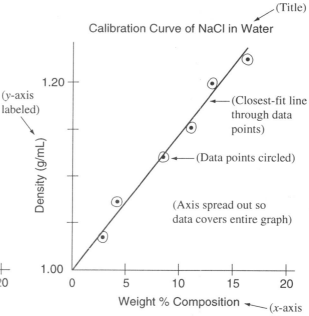

Calibration Curve of NaCl in Water

CORRECT. Best line through points and
dotted lines to determine value of unknown.

Experimental

Equipment: 10-mL graduated cylinder
100-mL graduated cylinder
balance
straight edge, or ruler

Materials: distilled water
a solid (examples: iron nail, copper penny, zinc, aluminum)
5% NaCl solution (sodium chloride)
10% NaCl solution
15% NaCl solution
20% NaCl solution

Procedure

Density of a Liquid

1. Determine and record the mass of a 10-mL *clean* and *dry* graduated cylinder.
2. Fill the 10-mL graduated cylinder halfway with distilled water. Read the volume using the bottom of the meniscus and the method shown in Figure 2.1. Record your volume to ±0.01 mL.

Figure 2.1
The proper method of reading a meniscus so that parallax error is avoided. The volume of liquid is 7.50 mL.

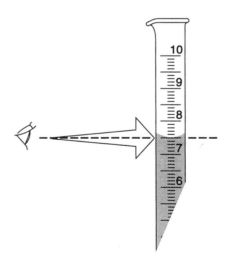

3. Determine and record the mass of your cylinder *with this volume of water in it.* The difference between the mass of the empty cylinder and the cylinder containing water is the mass of the water. Record this value.
4. Calculate the density of water. Record.

Density of a Solid

1. Obtain a solid from your instructor and record its identity.
2. Determine the mass of this solid and record.
3. Pour approximately 70 mL (it doesn't have to be exact) of distilled water into a 100-mL graduated cylinder. Read and record the actual volume, to one place after the decimal (±0.1 mL), by reading the bottom of the meniscus.
4. Carefully add the solid to the cylinder so that no water is lost. Read the new volume and record. The difference between the two readings is the volume of the solid.
5. Calculate the density of the solid.

Density of a Solution

NOTE: The concentration of a solution is sometimes described in terms of the solution's percentage composition on a weight basis. For example, a 5% NaCl solution contains 5 g of NaCl in every 100 g of solution (which corresponds to 5 g of NaCl for every 95 g of water present). Solutions of NaCl in distilled water consisting of the following weight percentages have been prepared: 5%, 10%, 15%, and 20%.

1. Clean and *dry* your 10-mL graduated cylinder. Determine its mass and record.
2. Fill the graduated cylinder approximately halfway full with the 5% NaCl solution. Read and record the precise volume to two places after the decimal.
3. Determine the mass of the graduated cylinder with the NaCl solution and record. The difference between this reading and the mass from step 1 is the mass of the NaCl solution. Record this value.
4. Calculate the density of the solution and record.
5. Repeat steps 1–4 for the 10%, 15%, and 20% NaCl solutions.
6. Construct a graph of the density of your solutions (on *y*-axis) versus the weight percentage composition (on *x*-axis). Be sure to include the pure water density value (0% NaCl) from the first Procedure, Density of a Liquid. The unknown should not be plotted until after the calibration curve has been drawn.
7. Obtain a sample of the unknown NaCl solution. Carefully determine the mass and actual volume of approximately 5 mL of the sample. Record these values.
8. Calculate the unknown solution's density and record. Using the graph from step 6, calculate the unknown's weight percentage composition. Record this value.

3
Heat of a Reaction From Calorimetry

Objective

- Measure the heat evolved or absorbed when hydrochloric acid reacts with sodium hydroxide.

Introduction

In general, all chemical reactions are accompanied by changes in energy, usually in the form of heat. A reaction that evolves heat is called an **exothermic reaction** and one that absorbs heat is called an **endothermic reaction.**

The **law of conservation of energy** states that energy is neither created nor destroyed in a chemical reaction, but is merely transformed from one form to another. In mathematical terms this can be stated as:

$$q_{\text{lost from reactants}} = -q_{\text{gained by surroundings}}$$

where q is the quantity of heat.

The heat evolved or absorbed during a reaction can be measured with a calorimeter. A *calorimeter* is a device for measuring the amount of heat evolved or absorbed in a chemical process. You will construct a calorimeter from some water in a plastic foam coffee cup. The heat change is then determined from the water temperature before and after the reaction occurs. In this experiment the total heat change during the reaction is equal to the heat change experienced by the solution (or water) and the calorimeter. In other words, the equation for heat flow becomes:

$$q_{\text{reaction}} = -(q_{\text{water}} + q_{\text{calorimeter}})$$

The following equations show how the heat gained or lost by the water (q_{water}) and the calorimeter ($q_{\text{calorimeter}}$) can be calculated:

$$q_{\text{water}} = (\text{sp. ht.}) \times (\text{mass of H}_2\text{O}) \times (t_f - t_i)$$

$$q_{\text{calorimeter}} = (C) \times (t_f - t_i)$$

where sp. ht. is the specific heat of water, C is the heat capacity of the calorimeter, t_f is the final temperature, and t_i is the initial temperature. Assume the specific heat of water is 4.18 J/(g · °C), the density of water is 1.00 g/mL, and the heat capacity of the calorimeter is 10.0 J/g°C.

These equations assume that the calorimeter is perfectly insulated. This assumption is not warranted since heat will flow through the walls. After all, hot coffee in a coffee cup will cool even if the top is covered. Because heat leaks through the walls, we will not be able to observe the highest (or lowest) temperature that could have been achieved in a perfectly insulated calorimeter. However, we need to know that temperature because it is t_f in our equations for heat flow.

We estimate t_f by plotting temperature as a function of time. We then extrapolate to the time at which the solutions were mixed. This temperature is t_f. A typical result is shown in Figure 3.1, which indicates how a final temperature of 31.8°C was estimated. Note, however, that the rate and method of stirring affects the appearance of the graph.

Figure 3.1
A graph showing temperature
as a function of time after
50 mL of a solution of NaOH
and 50 mL of a solution of
HCOOH are mixed.

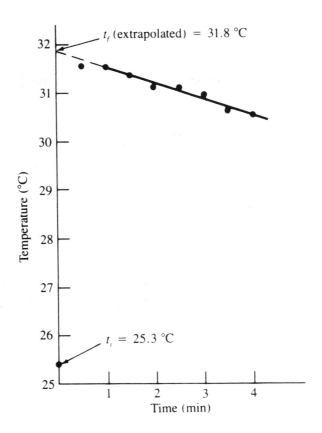

The chemical reaction that occurs during this experiment is

$$NaOH + HCl \longrightarrow NaCl + H_2O$$

You will measure the heat evolved or absorbed during this reaction.

Example Problem

Suppose the temperature of 50.0 mL of a soluti〜 〜H in a coffee-cup
calorimeter is 25.3°C. When 50.0 mL of a solut〜 〜rmic acid),
whose temperature is also 25.3°C, is added to 〜 〜ture
increases to 31.8°C (see Figure 3.1). The chemj

$$NaOH + HCOOH \longrightarrow \mathbb{N}$$

What is the heat of the reaction, $q_{reaction}$, a〜
place during the reaction?
    Solution: First let's calculate q_{water}.

$$q_{water} = (sp.\ ht.) \times$$

The total volume in the reaction is 50.0 mL (NaOH) plus 50.0 mL (HCOOH), for a final volume of 100.0 mL. The equation requires the total mass of water. Assuming the density of water is 1.00 g/mL, then:

$$(50.0 + 50.0) \text{ mL} \times 1.00 \text{ g/mL} = 100. \text{ g}$$

The specific heat of water is 4.18 J/(g · °C), t_f is 31.8°C, and t_i is 25.3°C.

$$q_{water} = (4.18 \text{ J/g} \cdot °C) \times (100. \text{ g}) \times (31.8°C - 25.3°C)$$

$$= 2.7 \times 10^3 \text{ J}$$

To calculate $q_{calorimeter}$, we must assume that the heat capacity of the calorimeter, C, is 10.0 J/°C.

$$q_{calorimeter} = (C) \times (t_f - t_i)$$

$$= (10.0 \text{ J/}°C) \times (31.8°C - 25.3°C)$$

$$= 65 \text{ J}$$

Then:

$$q_{reaction} = -q_{water} - q_{calorimeter}$$

$$= -2.7 \times 10^3 \text{ J} - 65 \text{ J}$$

$$= -2.8 \times 10^3 \text{ J, or } -2.8 \text{ kJ}$$

Experimental

Equipment: 2 6-ounce plastic foam coffee cups
plastic lid for one cup
stirring wire
400-mL beaker
2 100-mL graduated cylinders
1 thermometer

Materials: a solution of HCl (hydrochloric acid)
a solution of NaOH (sodium hydroxide)

Safety Precautions

Hydrochloric acid and sodium hydroxide can cause chemical burns in addition to ruining your clothing. If you spill one of these solutions on you, wash the contaminated area thoroughly and report the incident to your laboratory instructor. You may require further treatment.

Procedure

Making a Calorimeter

1. Work with a partner.
2. Nest two similarly sized plastic foam coffee cups. Obtain a tightly fitting plastic lid. To avoid tipping, support the calorimeter in a 400-mL beaker, as shown in Figure 3.2.
3. Using a narrow spatula, make two crosswise cuts in the lid, one near the center of the lid to insert the thermometer and one to the side to insert the stirring wire.
4. Insert the thermometer and stirring wire. Adjust the thermometer so that its bulb is about ½ inch above the bottom of the cup.

Measuring the Evolution or Absorption of Heat

1. Obtain exactly 50.0 mL of the HCl solution in a clean, dry graduated cylinder. Obtain exactly 50.0 mL of the solution of NaOH in another clean, dry graduated cylinder.
2. Measure the temperature of each solution, using the same thermometer, but rinse and dry it after the first measurement. If the temperatures are not within 0.5°C of one another, cool the warmer solution by immersing the graduated cylinder in tap water or warm the cooler solution with your hands. Record the average temperature. This is the initial temperature t_i.
3. Add the acid to the calorimeter.
4. Add the base to the calorimeter.
5. Immediately place the top on the calorimeter and begin stirring.
6. Record the temperature to the nearest 0.1°C after 30 s and every 30 s thereafter for 4 min.
7. On graph paper, plot the temperature against the time. Draw a straight line to extrapolate your results to the time of mixing (time = 0 s). Record the extrapolated temperature. This is the final temperature t_f.
8. Calculate $q_{reaction}$ using 4.18 J/g · °C for the specific heat of the solution, 1.00 g/mL for the density of the solution, and 10.0 J/°C for the heat capacity of the calorimeter.
9. Repeat steps 1 through 10 using a new piece of graph paper.
10. Calculate the average value of $q_{reaction}$.
11. Share your results with your classmates, and obtain results from as many of them as you can and determine the average.

Figure 3.2
The coffee-cup calorimeter.

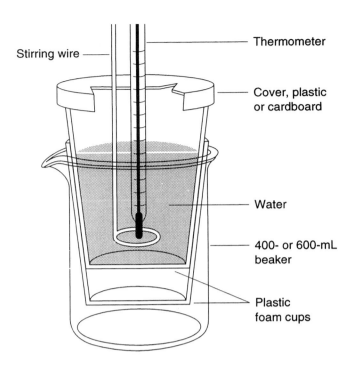

Stirring wire

Thermometer

Cover, plastic
or cardboard

Water

400- or 600-mL
beaker

Plastic
foam cups

Heat of a Reaction From Calorimetry

Prelaboratory Assignment

1. Give the chemical equation for the reaction that occurs during this experiment.

$$q = m \cdot \Delta t \cdot SH$$

2. a. When solutions of two reactants were mixed in a coffee-cup calorimeter, the following temperatures were recorded as a function of time. Plot the data on graph paper. Obtain t_f, the final temperature, by extrapolating with a straight line to the time of mixing (time = 0 s). The initial temperature, t_i, was 24.3°C.

Time (s)	t (°C)
30	38.6
60	39.7
90	39.5
120	39.4
150	39.3
180	39.2
210	39.1
240	39.0

t_f _____

b. Is this an exothermic or endothermic reaction? Why?

c. Why does the temperature increase, reach a maximum, and then decrease?

3. What special safety precautions must be observed during this experiment?

3
Heat of a Reaction
From Calorimetry

NAME _____

SECTION _____

DATE _____

Results

Reaction of HCl with NaOH

t_i (°C) _____

Temperature (°C) after

      30 s _____

      60 s _____

      90 s _____

     120 s _____

     150 s _____

     180 s _____

     210 s _____

     240 s _____

t_f (°C) _____

$q_{reaction}$ (J) _____

Show all calculations:

Pooled results (include your own)

Reaction	$q_{reaction}$				Average $q_{reaction}$ (kJ)
NaOH–HCl	_____	_____	_____	_____	
	_____	_____	_____	_____	
	_____	_____	_____	_____	
	_____	_____	_____	_____	_____

Questions

1. How many grams of water could be warmed from 25°C to 35°C using the
    average value of the heat that your class found in this experiment? (Show
    calculations)

2. How many grams of ice (specific heat = 2.06 J/g · °C) could be warmed from
    −12°C to −2°C using the average value of the heat that your class found in
    this experiment?

4

Separation of Cations
by Paper Chromatography

Objectives

- Separate and detect the component ions of a mixture by paper chromatography.

- Calculate the retention factor, R_f, for each ion in the mixture.

- Analyze a mixture of unknown cations by paper chromatography.

Introduction

In paper chromatography a drop of solution containing a mixture of substances is placed near one end of a rectangular piece of filter paper. The paper serves as the **stationary phase** (it does not "move" during the actual chromatography experiment). The end of the paper that is closer to the drop is immersed in a liquid to a point just below the drop. The liquid is the **mobile phase**. *Capillary action* (the same phenomenon that causes a liquid, such as coffee, to travel up a paper towel when the edge of the towel is immersed in a coffee spill) causes the liquid to flow up the chromatography paper. When the liquid reaches the spot, the components of the mixture begin to migrate upward with the mobile phase. Each component has a characteristic chemical affinity for the paper and a characteristic chemical affinity for the liquid. These affinities are competitive: the component's affinity for the paper tends to hold the component in one place, but its affinity for the liquid tends to make the component follow the liquid as it moves upward. A component with a strong affinity for the paper and a weak affinity for the liquid moves more slowly than a component with a weaker affinity for the paper and a stronger affinity for the liquid.

A substance's affinities for the stationary and mobile phases are entirely chracteristic of that substance. Different substances have their own characteristic affinities. Because each component of a mixture has its own characteristic affinities, each component travels up the paper at its own characteristic rate. If the paper is sufficiently large, all the components can be separated by the time the liquid has reached the top of the paper.

Each component now appears as a separate spot. If the components are highly colored, the spots will be visible. You can convert weakly colored or colorless spots to more highly colored ones by spraying them with substances that react with the components in the spots. The chromatography paper then contains a vertical array of colored spots arranged according to their characteristic rates of ascent. The word **chromatography**, which is derived from two Greek words and literally means "written in color," was coined to describe this phenomenon.

The distance traveled by a component of a spot with respect to the distance traveled by the mobile phase is a measure of that component's competitive affinities for the stationary and mobile phases. We define the component's R_f (**retention factor**) value in those terms:

$$R_f = \frac{\text{distance traveled by spot from origin}}{\text{distance from origin to solvent front}}$$

where the **origin** is the point at which the sample was originally placed on the paper, and the **solvent front** is the line representing the most advanced penetration of the paper by the mobile phase. The R_f value of a substance is characteristic of that substance and should be a constant under the same experimental conditions.

In this experiment you will separate a mixture of iron(III), copper(II), and nickel(II) ions, Fe^{3+}, Cu^{2+}, and Ni^{2+}, respectively, by paper chromatography. Iron(III) ions are typically colored reddish-brown; copper(II) and nickel(II) are blue and green but may be hard to observe here. You will spray the chromatography paper after the experiment with aqueous ammonia (which reacts with copper) and dimethylglyoxime (which reacts with nickel) to form colored complexes.

A Special Note: The labels on some of the bottles used in this experiment may include a statement about the concentration of the solution in the bottle: 1.0 M. The symbol M stands for molarity with units of mol/L, and it is a measure of how many molecules or ions are in each liter of solution. You will become much more familiar with this unit in subsequent experiments.

Experimental

Equipment: 800- or 1000-mL beaker
100-mL graduated cylinder
plastic film (e.g., Saran wrap)
chromatography paper
pencil
rubber band
straight edge or ruler
6 capillary tubes
heat lamp
stapler

Materials: 1.0 M $CuSO_4$ [copper(II) sulfate]
1.0 M $FeCl_3$ [iron(III) chloride]
1.0 M $NiSO_4$ [nickel(II) sulfate]
acetone
concentrated HCl (hydrochloric acid)
concentrated $NH_3(aq)$ (aqueous ammonia)
1% dimethylglyoxime in alcohol
2 unknown mixtures

Safety Precautions

Acetone is extremely flammable. Make sure there are no open flames in the vicinity of your work area. Fumes of acetone, concentrated hydrochloric acid, and ammonia are all objectionable and, to some degree, harmful. These chemicals should be used only in the hood. Hydrochloric acid and ammonia can cause severe chemical burns in addition to ruining your clothing. If you spill one of these substances on you, wash the contaminated area thoroughly and report the incident to your instructor. You may require further treatment.

Figure 4.1
A piece of chromatography paper properly prepared for use in this experiment.

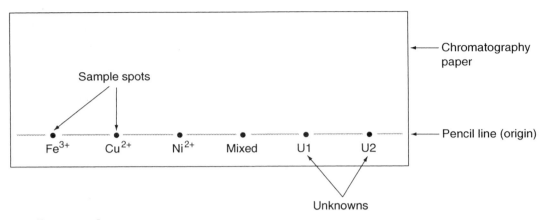

Procedure

Getting Started

1. Obtain your 2 unknown mixtures, 6 capillary tubes, chromatography paper, a piece of clear plastic film, a rubber band, and a 800-mL beaker.

2. Using a graduated cylinder and working in the hood, add 19 mL acetone, 2 mL distilled water, and 4 mL concentrated HCl to the 800-mL beaker. Cover the beaker with the plastic film, and hold the film in place with a rubber band. This allows the atmosphere within the beaker to become saturated with solvent vapor and helps to give a better chromatographic separation.

3. Place the chromatography paper on a clean paper towel. Using a pencil, draw a line about 1 cm from the long edge of the paper. This line will indicate the origin (see Figure 4.1). Use an ordinary lead pencil for this line. Ink ofter contains substances that may be soluble in the solvent, producing chromatograms of their own.

4. Using a different capillary tube for each solution, transfer a drop of each listed solution to the pencil line, as shown in Figure 4.1. The smaller you make the spots, the better the chromatogram. You may want to practice on a piece of scrap paper to get the spots small. Space the spots evenly, and leave a margin of about 3 cm from the short edge of the paper. With a pencil label each spot by writing on the paper directly beneath the spot. The solutions are:
   a. Fe^{3+} solution
   b. Cu^{2+} solution
   c. Ni^{2+} solution
   d. solution of all three ions (Fe^{3+}, Cu^{2+}, Ni^{2+})
   e. 2 different unknown solutions (be sure to record their identificatio: numbers)

5. Dry the paper under a heat lamp. Form the paper into a cylinder withou overlapping the edges. Fasten the paper with staples, as shown in Figure 4.:

Figure 4.2
The experimental arrangement, showing the covered beaker containing the stapled chromatography paper (the stationary phase) and the solution (the mobile phase).

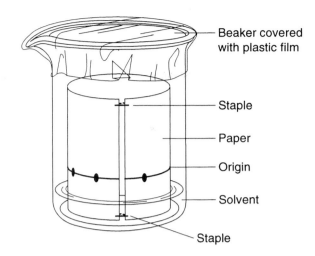

Beaker covered with plastic film

Staple

Paper

Origin

Solvent

Staple

Doing the Experiment

1. Remove the plastic film from the 800-mL beaker, and gently place the paper cylinder inside. The origin line should be above the surface of the liquid. Be careful that the paper does not touch the inside wall of the beaker. Do not splash the solution in the beaker. Replace the plastic film. (See Figure 4.2.)
2. The beaker must be absolutely stationary throughout the experiment. To view the paper in its entirety, move to another position rather than turning the beaker.
3. Allow the liquid to ascend to within 2–3 cm from the top of the paper (about 30–40 min will be required). You should be able to observe the progress of some of the spots during the ascent.
4. When the liquid has reached the desired height, remove the paper from the beaker. Place the wet paper on a paper towel, and trace with a pencil the solvent front position. Remove the staples and dry the paper under a heat lamp.
5. Wearing rubber or plastic gloves and working in a hood, spray the paper with an NH_3 solution. The paper should be moist but not wet. Next spray the paper with a solution of dimethylglyoxime. A dark spot should now be visible for each substance. Dry the paper.
6. Circle each spot with a pencil. Measure the vertical distance that the approximate center of each spot has traveled from the origin. Your measurement should be to within ±0.1 cm. Record your results. Also record the color of each spot.
7. Measure and record the vertical distance from the origin to the solvent front position for each spot. Calculate the R_f value for each spot.
8. Use the R_f values to deduce the identity of each component in your unknown mixtures.
9. Attach the dry chromatography paper to your report.

4
Separation of Cations
by Paper Chromatography

NAME _____

SECTION _____

DATE _____

Prelaboratory Assignment

1. Why must a pencil rather than a pen be used for marking a chromatogram?

2. Why should you avoid moving the beaker while the chromatogram is being developed?

3. What special safety precautions must be taken during this experiment?

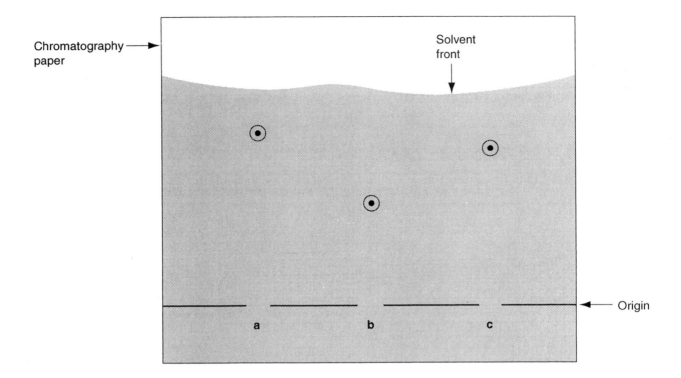

4. A sample chromatogram is shown in the figure above. Find the R_f values for the spots shown. Using a metric ruler, measure to a tenth of a centimeter (i.e., a millimeter). Measure the distance from the origin line to the middle of the spot in the circle. Show all calculations below.

    a. $R_f =$

    b. $R_f =$

    c. $R_f =$

4
Separation of Cations by Paper Chromatography

NAME _____

SECTION _____

DATE _____

Results

Solution	Ion	Color	Distance (mm) from origin to		R_f
			Solvent front	Spot center	
Fe^{3+}	Fe^{3+}				
Cu^{2+}	Cu^{2+}				
Ni^{2+}	Ni^{2+}				
Mixed	Fe^{3+}				
	Cu^{2+}				
	Ni^{2+}				

Unknown no. _____

Ions present _____

Unknown no. _____

Ions present _____

Questions

1. Why must the container be covered during chromatography?

2. Why must the solvent level be below the spot of solution when the chromatogram is placed in the elution tank?

3. Why must elution of the solutions be stopped before the solvent front rises to the top of the paper?

4. What is an R_f value?

5. What would happen if the solvent traveled only halfway up the filter paper? Would the experiment still work?

5
Qualitative Analysis for Chromium, Iron, and Copper

Objectives

- Separate a mixture of Cr^{3+}, Fe^{3+}, and Cu^{2+}.

- Identify each ion in the mixture.

- Determine which of these ions are present in an unknown mixture.

Introduction

Qualitative analysis answers the question: what compounds or ions are present in a particular unknown solution? The next step is often **quantitative analysis,** which answers the question: how much of a specific compound or ion is present? In this experiment you will analyze a solution containing three transition metal ions: Cr^{3+}, Fe^{3+}, and Cu^{2+}. Typically, a mixture of metal cations must be separated before they can be identified. The cations are identified by adding a series of reagents that cause each ion present to react in a certain way.

Separation begins when aqueous ammonia, $NH_3(aq)$, is added to a mixture of the three ions. Aqueous ammonia is a source of both ammonia, NH_3, and hydroxide ions, OH^-. As a consequence, this reagent causes the precipitation of the hydroxides of Cr^{3+} and Fe^{3+}, as shown in the following net ionic reaction for Cr^{3+}.

$$Cr^{3+}(aq) + 3OH^-(aq) \longrightarrow Cr(OH)_3(s)$$

Aqueous ammonia also causes the formation of $Cu(NH_3)_4^{2+}$, a deep blue complex ion, which remains in solution. Decanting (pouring off) the solution from the precipitates separates copper from chromium and iron. The presence of copper in the solution is confirmed when a red-maroon precipitate of $Cu_2Fe(CN)_6$ forms after the addition of $K_4Fe(CN)_6$.

To continue the separation, the hydroxides of Cr^{3+} and Fe^{3+} are treated with hydrogen peroxide, H_2O_2, which causes the $Cr(OH)_3$ precipitate to become the soluble yellow chromate ion, CrO_4^{2-}. The remaining precipitate of $Fe(OH)_3$ is separated from the chromate ion by decanting the solution. The presence of chromium in the solution is confirmed when a yellow precipitate of $PbCrO_4$ forms after a solution a $Pb(NO_3)_2$ is added.

The $Fe(OH)_3$ precipitate is then dissolved in acid. The presence of iron is confirmed by the formation of a deep red color due to $Fe(SCN)^{2+}$, a complex ion, after the addition of a solution of KSCN.

Figure 5.1 provides a scheme for qualitative anaylsis of the three metal cations. You can use this scheme to separate and identify any combination of these metal ions in an unknown mixture. Also note that each of these metal ions has a characteristic color. You can use the color of your unknown mixture to support your anaylsis using this scheme.

Experimental

Equipment: 6 small (13 mm × 100 mm) test tubes
stirring rods
centrifuge
plastic wash bottle filled with distilled water
blue litmus paper
Bunsen burner

Figure 5.1
A qualitative analysis scheme for separating and identifying Cr^{3+}, Fe^{3+}, and Cu^{2+}.

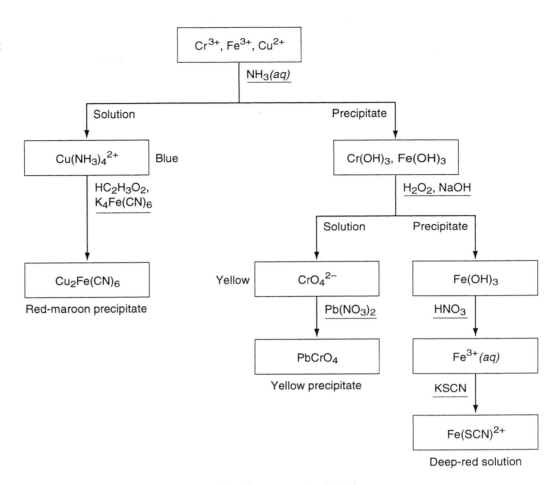

Materials:

test tube holder
dropper bottles

6.0 M $NH_3(aq)$ [aqueous ammonia, also known as ammonium hydroxide, $NH_4OH(aq)$]
3% H_2O_2 (hydrogen peroxide)
6.0 M NaOH (sodium hydroxide)
6.0 M HNO_3 (nitric acid)
0.1 M KSCN (potassium thiocyanate)
6.0 M $HC_2H_3O_2$ (acetic acid)
0.1 M $K_4Fe(CN)_6$ (potassium ferrocyanide)
0.1 M $Pb(NO_3)_2$ (lead nitrate)
0.1 M $Cr(NO_3)_3$ (chronium(III) nitrate; solution A)
0.1 M $Fe(NO_3)_3$ [iron(III) nitrate; solution B]
0.1 M $Cu(NO_3)_2$ (copper(II) nitrate; solution C)
unknown solution

Safety Precautions

Never attempt to stop the centrifuge rotor with your fingers!

Aqueous ammonia, sodium hydroxide, nitric acid, and acetic acid can cause chemical burns in addition to ruining your clothes. If you spill any of these on you, wash the contaminated area thoroughly with tap water, and report the incident to your instructor. You may require further treatment.

Before you leave the laboratory, make sure that your gas outlet and those of your neighbors are closed. Finally, wash your hands. Solutions containing lead are poisonous.

Procedure

Getting Started

1. Obtain 6 small test tubes.
2. Obtain your unknown solution and record its identification number and color.
3. Prepare a known mixture of the three ions, using 5 mL of 0.1 M $Cr(NO_3)_3$, 5 mL of 0.1 M $Fe(NO_3)_3$, and 5 mL of 0.1 M $Cu(NO_3)_2$. Note and record the color of each solution before you mix them together. Make sure that the final solution is thoroughly mixed.
4. Conduct the analysis of the known and unknown solutions simultaneously so that you can compare the results.
5. Label all test tubes so that you do not confuse the known and unknown solutions and precipitates at any time.
6. Obtain instructions for using the centrifuges in your laboratory.
7. Obtain instructions for discarding the solutions you will use in this experiment.
8. Take care in handling the solutions used in this experiment. Remember to wash your hands thoroughly if you come in contact with any harmful chemical or one whose identification is unknown.

Doing the Analysis

1. Carefully pour 1 mL of the known mixture and 1 mL of the unknown mixture into separate test tubes.
2. Each subsequent addition and operation should be conducted on both mixtures.
3. Add 20 drops of 6.0 M $NH_3(aq)$ and stir with a clean stirring rod.
4. If no precipitate forms, proceed with step 12. If a precipitate is present, centrifuge the mixture for about 1 min and decant (pour off) the solution. Save the solution for step 12, and use the precipitate in the next step.
5. Spray the precipitate in the test tube with about 1 mL of distilled water from a plastic wash bottle. Stir with a clean stirring rod and centrifuge the mixture. Discard the water, and use the precipitate in the following step.
6. Add 12 drops of 3% H_2O_2, 1 mL of distilled water, and 5 drops of 6.0 M NaOH to the precipitate. Let this mixture stand for about 1 min.

7. Light a Bunsen burner. Heat the mixture gently and cautiously to decompose the remaining H_2O_2. Use a test tube holder. Do not let the flame linger in any one place so as to avoid "bumping" (an eruption of bubbles due to boiling quickly). Heat until effervescence (bubbling) from the decomposition ceases.

8. Cool the test tube briefly under tap water. Centrifuge the mixture for 2 minutes, or longer if separation is incomplete. Save the solution in a clean test tube for step 13. If a precipitate remains, use it in the following step. If no precipitate remains, proceed to step 13.

9. Wash the precipitate in the same manner as in step 5.

10. Dissolve the precipitate with 5 drops of 6.0 M HNO_3. Add 3 mL of distilled water, and mix with a glass rod to obtain a homogeneous solution.

11. Add 5 drops of 0.1 M KSCN to this solution and mix thoroughly. A deep red color confirms the presence of iron.

12. Make the solution from step 4 acidic by adding drops of 6.0 M $HC_2H_3O_2$ until the solution turns blue litmus paper pink. Add 10 drops of 0.1 M $K_4Fe(CN)_6$ and mix thoroughly. A red-maroon precipitate confirms the presence of copper.

13. If you did step 8, make that solution acidic by adding drops of 6.0 M $HC_2H_3O_2$ until the solution turns blue litmus paper pink. Add 10 drops of 0.1 M $Pb(NO_3)_2$ and mix thoroughly. Centrifuge the mixture. A yellow precipitate confirms the presence of chromium.

14. Record the ions that are present in the unknown mixture. Is the color of the unknown mixture in accordance with your conclusions? If not, repeat the analysis.

CAUTION: Before you leave the laboratory, make sure that your gas outlet and those of your neighbors are closed. Finally, wash your hands. Solutions containing lead are poisonous.

5
Qualitative Analysis for Chromium, Iron, and Copper

NAME _____

SECTION _____

DATE _____

Prelaboratory Assignment

1. How will the color of the unknown provide a clue about the presence or absence of one of the possible components of the unknown mixture?

2. Write the three balanced chemical equations that describe the formation of the following three precipitates: $Cr(OH)_3(s)$, $Fe(OH)_3(s)$, and $PbCrO_4(s)$. Figure 5.1 may be helpful.

3. What special safety precautions are cited in this experiment?

5
Qualitative Analysis for
Chromium, Iron, and Copper

NAME _____

SECTION _____

DATE _____

Results

Unknown no. _____

Color of unknown mixture: _____

Color of $Cr(NO_3)_3$ solution: _____

Color of $Fe(NO_3)_3$ solution: _____

Color of $Cu(NO_3)_2$ solution: _____

Ions present in unknown mixture: _____

Questions

1. To decompose excess H_2O_2 during this experiment, you heated the solution until effervescence ceased. The effervescence was due to the evolution of O_2. No other gas was evolved. The other product was water. Write the balanced chemical equation that describes this decomposition.

2. Suppose that a solution contained only one cation, either Cr^{3+}, Fe^{3+}, or Cu^{2+}. What would you do to identify the cation quickly?

6
Ionic Reactions in Aqueous Solutions

Objectives

- Study the nature of ionic reactions.
- Write the net ionic equations for a series of precipitation reactions.

Introduction

Because of its high polarity, water is a particularly excellent solvent for ionic compounds. When ionic compounds are dissolved in water, the ions separate and become surrounded by water molecules, as shown in Figure 6.1. The separation of ions is known as **dissociation**.

Some substances are not very soluble in water. Frequently such materials are generated *in situ* by mixing various other substances that are very soluble in water. For example, silver chloride (AgCl) is not soluble in water. If an aqueous solution of silver nitrate ($AgNO_3$) (very soluble) is mixed with an aqueous solution of sodium chloride (NaCl) (very soluble), the combination of silver cations (Ag^+) from one solution and chloride (Cl^-) anions from the other solution generate silver chloride (AgCl), which precipitates and settles to the bottom of the container. If you did Experiment 1, you might recall performing this precipitation reaction. An example of the formation of a precipitate is shown in Experiment 1, Figure 1.1. The solution that remains above the AgCl precipitate is essentially a solution of sodium nitrate ($NaNO_3$). Silver ions and sodium ions have switched partners, ending up in a compound with the anion that originally came from the other substance:

$$AgNO_3(aq) + NaCl(aq) \longrightarrow AgCl(s) + NaNO_3(aq)$$

This precipitation reaction is an example of a double-replacement reaction.

To clarify what is really happening in these precipitation reactions, it is often more instructive to write the reaction in the net ionic form. The net ionic reaction includes only the ions involved in actually forming the precipitate. The other

Figure 6.1
When ionic compounds are dissolved in water, the ions separate from one another and are surrounded by water molecules.

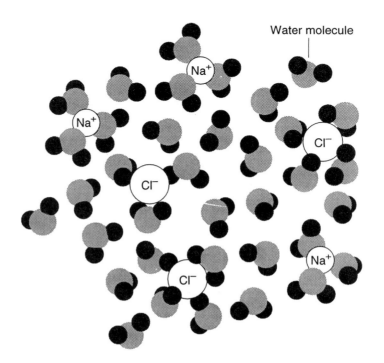

ions are called the *spectator ions,* and they are present in the solution after the precipitate has formed. For example, for the reaction between $AgNO_3$ and $NaCl$, sodium and nitrate are the spectator ions, and the net ionic reaction is:

$$Ag^+(aq) + Cl^-(aq) \longrightarrow AgCl(s)$$

The net ionic reaction is especially instructive; for example, it implies that any solution containing silver cations will react with any solution containing chloride anions to form AgCl. If we mixed an aqueous solution of $AgNO_3$ with an aqueous solution of KCl, instead of NaCl, we would again expect a precipitate of AgCl to form, and the net ionic reaction would be identical to the one with NaCl.

Some general guidelines for writing a net ionic equation are as follows:

1. List the formula of the precipitate formed as a product in the reaction. Place a (s), for solid (insoluble in water), after the formula for the product. Make sure the formula is written correctly. Write the cation first and the anion last. The charge of the cation becomes the subscript on the anion, and vice versa. Use parentheses if polyatomic ions are involved. Simplify the formula if the subscripts are divisible by the same small number, as in $BaCO_3$. For example:

sodium sulfide	Na^+	S^{2-}	$= Na_2S(aq)$
calcium hydroxide	Ca^{2+}	OH^-	$= Ca(OH)_2(s)$
iron(III) nitrate	Fe^{3+}	NO_3^-	$= Fe(NO_3)_3(aq)$
barium carbonate	Ba^{2+}	CO_3^{2-}	$= BaCO_3(s)$

    We might begin writing the net ionic equation for the precipitation reaction that generates silver carbonate from silver nitrate and sodium carbonate as follows:

    $$\longrightarrow Ag_2CO_3(s)$$

2. List the component ions that make up this formula as reactants. Make sure the charges are correct. See Table 6.1 for the names and formulas of the cations and anions used in this experiment. Write (aq), for aqueous (soluble in water), after the cation and anion.

    To continue our example:

    $$Ag^+(aq) + CO_3^{2-}(aq) \longrightarrow Ag_2CO_3(s)$$

3. Balance the equation by adjusting the numbers in front of the cation and anion so that they correspond with the subscripts in the formula on the right of the arrow.

    $$2Ag^+(aq) + CO_3^{2-}(aq) \longrightarrow Ag_2CO_3(s)$$

Table 6.1

Names and Formulas of Ions Used in Experiment 6

Cations

+3 Ions	+2 Ions	+1 Ions
Al^{3+}, aluminum	Ba^{2+}, barium	Na^+, sodium
Fe^{3+}, iron(III)	Sr^{2+}, strontium	K^+, potassium
	Co^{2+}, cobalt	Ag^+, silver
	Ni^{2+}, nickel	
	Mg^{2+}, magnesium	

Anions

	−2 Ions	−1 Ions
	CO_3^{2-}, carbonate	Cl^-, chloride
	SO_4^{2-}, sulfate	OH^-, hydroxide
		NO_3^-, nitrate

In this experiment you will be asked to use the data from several reactions to determine which combinations result in the formation of precipitates. Some very important principles will help you determine the precipitate:

1. Once a compound is determined to be soluble in water, it will always be soluble in water.
2. Once a compound is determined to be a precipitate (insoluble in water), that combination of ions will always form a precipitate.
3. When two solutions are mixed and no precipitate forms, then each of the two new combinations of ions will also be soluble in water. For example: no precipitate forms when solutions of sodium chloride and potassium nitrate are mixed:

molecular equation $NaCl(aq) + KNO_3(aq) \longrightarrow NaNO_3(aq) + KCl(aq)$

net ionic equation $\longrightarrow$ (there is no net change)

From this we can conclude that $NaCl$, KNO_3, $NaNO_3$, and KCl are all soluble in water and will never be observed as a precipitate. This important result can be used in this experiment.

Experimental

Equipment: clear acetate sheet
micropipets

Materials: 0.1 M barium nitrate
0.1 M barium chloride
0.1 M barium hydroxide
0.1 M potassium chloride
0.1 M potassium hydroxide
0.1 M potassium sulfate
0.1 M sodium sulfate
0.1 M sodium hydroxide
0.1 M sodium nitrate
0.1 M silver nitrate
0.1 M aluminum sulfate
0.1 M strontium nitrate
0.1 M iron(III) chloride
0.1 M cobalt chloride
0.1 M magnesium sulfate
0.1 M magnesium chloride
0.1 M nickel chloride

Safety Precautions

Salts of the heavy metals are toxic. Many are harmful to the skin. Wash your hands immediately if any of these reagents get on your skin.

Procedure

1. In the following reactions, use a clean acetate sheet as a reaction chamber and several micropipets to dispense the reactants. Combine the reagent solutions as indicated in the list of reactants that follows by placing one drop of each of the two reactants together on the acetate sheet. Note whether or not a precipitate forms for each combination of reactants (the mixture will look cloudy). Record also the color of the solutions used and of the precipitates that form. *Use a new micropipet for every solution. Do not transfer the micropipets between different solutions.*
2. Rinse acetate sheets with distilled water. Be careful to collect the rinse water in a beaker to dispose of it later in the appropriate waste receptacle.
3. After all of the reactions have been completed, determine the formula of each precipitate formed, and write a net ionic equation for each reaction. Remember that a given combination of two ions can be eliminated as a precipitate if the combination is a reactant in another reaction, or if the two ions were mixed together in one of the mixtures that did not result in a

precipitate. In any of the reactions where a precipitate forms, only one of the products is the precipitate; the other is soluble in water.

Reactants

barium nitrate and potassium chloride $\longrightarrow$
barium nitrate and sodium sulfate $\longrightarrow$
barium nitrate and aluminum sulfate $\longrightarrow$
barium chloride and silver nitrate $\longrightarrow$
strontium nitrate and aluminum sulfate $\longrightarrow$
iron(III) chloride and potassium hydroxide $\longrightarrow$
cobalt chloride and potassium hydroxide $\longrightarrow$
cobalt chloride and sodium nitrate $\longrightarrow$
magnesium sulfate and sodium hydroxide $\longrightarrow$
barium hydroxide and nickel chloride $\longrightarrow$
barium hydroxide and magnesium chloride $\longrightarrow$
barium hydroxide and potassium sulfate $\longrightarrow$
sodium hydroxide and nickel(II) chloride $\longrightarrow$
sodium hydroxide and magnesium chloride $\longrightarrow$

6
Ionic Reactions in Aqueous Solutions

NAME _____

SECTION _____

DATE _____

Prelaboratory Assignment

1. Give the formulas for the component cations and anions and for the compounds.

Name	Cation	Anion	Compound
Potassium chloride			
Aluminum hydroxide			
Iron(III) sulfate			
Magnesium nitrate			

2. No precipitate forms when solutions of potassium chloride and sodium nitrate are mixed. What can you conclude about the solubilities of potassium chloride, sodium nitrate, potassium nitrate, and sodium chloride? Write a formula for each of these four substances.

3. When solutions of silver nitrate and potassium chloride are mixed, a white precipitate forms. Determine the correct net ionic equation using the information from Question 2.

6
Ionic Reactions in Aqueous Solutions

NAME ________________________________

SECTION ________________________________

DATE ________________________________

Results

Observations on mixing: include the color and formula of any precipitates formed.

Barium nitrate and potassium chloride $\longrightarrow$

Barium nitrate and sodium sulfate $\longrightarrow$

Barium nitrate and aluminum sulfate $\longrightarrow$

Barium chloride and silver nitrate $\longrightarrow$

Strontium nitrate and aluminum sulfate $\longrightarrow$

Iron(III) chloride and potassium hydroxide $\longrightarrow$

Cobalt chloride and potassium hydroxide $\longrightarrow$

Cobalt chloride and sodium nitrate $\longrightarrow$

Magnesium sulfate and sodium hydroxide $\longrightarrow$

Barium hydroxide and nickel chloride $\longrightarrow$

Barium hydroxide and magnesium chloride $\longrightarrow$

Barium hydroxide and potassium sulfate $\longrightarrow$

Sodium hydroxide and nickel chloride $\longrightarrow$

Sodium hydroxide and magnesium chloride $\longrightarrow$

Question

1. Write the balanced net ionic equation for each reaction in this experiment that results in the formation of a precipitate.

7

The Empirical Formula of a Chemical Compound

Objectives

- Prepare a compound from its elements, and determine the mass of each element in the compound.

- Calculate the empirical formula of the compound.

Introduction

All chemical compounds are composed of atoms of different elements bonded together. The **empirical formula**, also known as the *simplest formula*, represents the smallest whole-number ratio of atoms in a compound. For example, the chemical compound water is composed of two hydrogen atoms for every one oxygen atom and has the empirical formula H_2O. The **molecular formula** represents the actual number of atoms in each molecule of a compound. The molecular formula may be identical with the empirical formula (as in water), or it may be some integral multiple of the empirical formula. For example, the molecule glucose (sugar) has the molecular formula $C_6H_{12}O_6$ and the empirical formula CH_2O.

To find the empirical formula of the compound in this experiment, you will combine the elements that make up the compound under conditions that will allow you to determine independently the mass of each element. The number of moles of each element in the compound is calculated by dividing the experimental values of mass by their corresponding known atomic weight, obtained from a periodic table. You then obtain the empirical formula by dividing the number of moles of each element by the smallest number of moles to convert the smallest number to 1. If all the numbers obtained are whole numbers, these numbers are the subscripts in the empirical formula. If all the numbers are not whole numbers, you must multiply the number of moles of each element by the smallest number possible such that all fractions are converted to whole numbers.

The synthesis of the compound used in this experiment involves the reaction of the element magnesium with an excess of the element oxygen from the air. This is accomplished by igniting the magnesium in air and letting it burn. After a known quantity of magnesium is burned, the product consists of magnesium oxide and a small amount of magnesium nitride, Mg_3N_2. Reacting magnesium nitride with water converts it to magnesium hydroxide, $Mg(OH)_2$, and ammonia, NH_3. Heating the magnesium hydroxide converts it to magnesium oxide and steam. By measuring the difference in mass of the starting material (pure magnesium) and the final product (magnesium oxide), you determine the mass of oxygen in the compound. Finally, by converting the masses into numbers of moles, you obtain the empirical formula of the compound.

Sample Calculations

A sample of chromium is ignited in a crucible. We will calculate the empirical formula of the oxide formed using the following data:

Average mass of empty crucible and lid	19.24 g
Mass of crucible, lid, and chromium	19.87 g
Average mass of crucible, lid, and oxide	20.16 g

1. Mass of chromium and oxygen:

Mass of crucible, lid, and chromium	−	mass of crucible and lid	=	mass of chromium
19.87 g	−	19.24 g	=	0.63 g

Mass of crucible, lid, and oxide	−	mass of crucible, lid, and chromium	=	mass of oxygen
20.16 g	−	19.87 g	=	0.29 g

2. Moles of chromium and of oxygen:

$$0.63 \text{ g Cr} \times \frac{1 \text{ mol Cr}}{52.00 \text{ g}} = 0.012 \text{ mol Cr atoms}$$

$$0.29 \text{ g O} \times \frac{1 \text{ mol O}}{16.00 \text{ g}} = 0.018 \text{ mol O atoms}$$

3. Empirical formula: Obtain the ratio of atoms by dividing the number of moles of each atom by the smallest number of moles:

$$\text{Cr:} \quad \frac{0.012 \text{ mol}}{0.012} = 1.0 \qquad \text{O:} \quad \frac{0.018 \text{ mol}}{0.012} = 1.5$$

The ratio is 1.0 mol chromium atoms to 1.5 mol oxygen atoms. Change the ratio to a whole-number ratio by multiplying each value by 2:

moles of Cr atoms	=	1.0×2	=	2.0
moles of O atoms	=	1.5×2	=	3.0

The empirical formula is therefore Cr_2O_3.

Experimental

Equipment: crucible and lid
clay triangle
iron ring and stand
Bunsen burner
crucible tongs
wire gauze
sandpaper

Materials: magnesium ribbon

Safety Precautions

Special care must be taken in operating a Bunsen burner and in the handling of hot objects. Tie long hair back so that it does not get into the flame. Be sure to wear safety goggles throughout the experiment. Handle the crucible and its lid only with crucible tongs. The blue flame from the Bunsen burner may be invisible at times, so be careful not to reach over a Bunsen burner unless you are absolutely sure that there is no flame.

Procedure

Getting Started

1. Obtain about 0.2 g of magnesium ribbon. (The weight need only be approximate; you will obtain a more precise value in step 2 under "Preparation of Magnesium Oxide.") If the surface is not shiny and bright, clean it with sandpaper.
2. Wash, rinse, and dry a crucible and lid.

Drying the Crucible and Lid

1. Support the crucible and lid on a clay triangle, as shown in Figure 7.1. Adjust the height of the iron ring so that the bottom of the crucible will be in the hottest part of the Bunsen burner flame. (The Introduction to this manual discusses Bunsen burners and their use.)
2. Heat the covered, empty crucible in the direct flame for approximately 4 min. The bottom of the crucible should become red-hot during this time.
3. Turn the burner off and move it away. Allow the crucible to cool for 5 to 10 min.

Figure 7.1
Correct arrangement of the ring stand, clay triangle, crucible with its lid, and Bunsen burner.

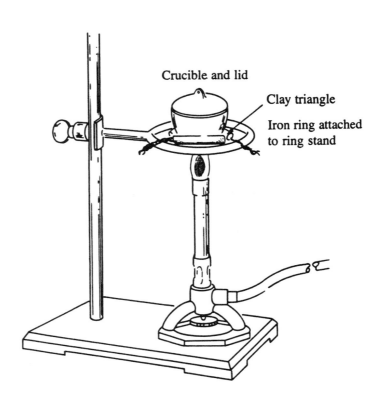

Crucible and lid

Clay triangle

Iron ring attached to ring stand

Figure 7.2
The proper method
for carrying the crucible,
with crucible tongs,
over wire gauze.
A watch glass
may be used
instead of the wire gauze.

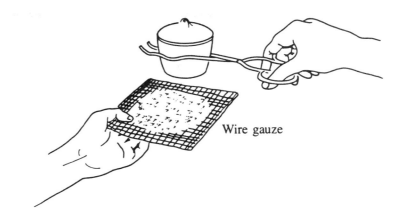

Wire gauze

4. The crucible is cool when you can place a finger about ½ in. from the bottom of the crucible and feel no heat. *Important! Do not touch the crucible directly with your fingers during the rest of this experiment!* The weight of a single fingerprint is enough to introduce significant error into your results!

5. Using crucible tongs, carefully carry the covered crucible to a balance. Hold a wire gauze under the crucible, as shown in Figure 7.2. Do not place the crucible and lid directly on the benchtop. Put them on the wire gauze or leave them in the clay triangle while waiting for a balance.

6. Obtain and record the mass of the covered crucible.

7. Repeat steps 1–6 until two consective mass readings do not differ by more than ±0.001 g (or by the precision suggested by your instructor). Record the average of these two final mass values.

Preparation of Magnesium Oxide

1. Place the *loosely* folded magnesium ribbon into the crucible. All of the ribbon must fit within the crucible. Replace the crucible lid.

2. Weigh the crucible, lid, and metal on a balance and record the mass.

3. Return the covered crucible to the clay triangle, using tongs and the wire gauze. Begin by gently heating the crucible by brushing the bottom of the crucible with the burner flame for about 2 or 3 min. Next, heat the crucible strongly by placing the burner back on the ring stand and heating the crucible in the hottest part of the flame for another 3 min.

4. Vent the crucible slightly by carefully sliding the lid with the crucible tongs. Do not open the lid too far or the magnesium itself will ignite. Minimize the amount of smoke produced. (The white smoke is made up of very finely divided magnesium oxide particles, and any that escape will cause an error in your final mass value.) The metal should glow brightly but not burst into flame during this procedure.

5. Repeat step 4 every few minutes until no metal is evident and no glow occurs when the lid is lifted.

6. Allow the covered crucible and its contents to cool. The contents should be grayish-white.

7. Remove the lid with the tongs and place it on the wire gauze. Add a few drops of distilled water directly to the contents in the crucible. The smell of NH_3 may be evident at this point.

8. Re-cover the crucible so that the lid is slightly ajar. Heat the crucible gently by brushing it with the flame until the water boils away and the contents are dry. Next, heat the crucible strongly for 8–10 min to convert the magnesium hydroxide to magnesium oxide.

9. Allow the covered crucible and its contents to cool to the same point as in step 6.

10. Obtain the mass of the covered crucible and record it.

11. Heat the covered crucible strongly again for about 3 min. Obtain and record the mass after the crucible is cool.

12. Repeat step 11 until two consecutive masses differ by no more than ±0.001 g.

13. Discard the magnesium oxide as specified by your instructor. Clean the crucible and lid carefully and thoroughly, and return all equipment to its original place.

Before leaving the laboratory make sure that your gas outlet and those around you are closed!

The Empirical Formula
of a Chemical Compound

NAME _____

SECTION _____

DATE _____

Prelaboratory Assignment

1. Which of the following is an empirical formula?

   a. $(NO_2)_2$ b. $C_{2.5}H_6O_{1.5}$ c. Na_3PO_4 d. C_6H_6

2. Write the balanced equation for the reaction between magnesium nitride and water.

3. A student reacted 0.672 g of iron with oxygen from the air. The final product weighed 0.961 g. Determine the empirical formula of the product, iron oxide.

7
The Empirical Formula
of a Chemical Compound

NAME _____

SECTION _____

DATE _____

Results

Mass of crucible and lid (g)   _____   _____

_____   _____

_____   _____

Average mass of crucible and lid (g)   _____

Mass of crucible, lid, and magnesium (g)   _____

Mass of magnesium (g)   _____

Mass of crucible, lid, and magnesium
    oxide (g)   _____   _____

_____   _____

Average mass of crucible, lid,
    and oxide (g)   _____

Average mass of crucible and lid (g)
    (from above)   _____

Mass of magnesium oxide (g)   _____

Questions (Show all work!)

1. What is the *mass* of the oxygen in your compound? What is the number of *moles* of oxygen in your compound?

2. How many *moles* of magnesium are in your compound?

3. What is the ratio of moles of magnesium to moles of oxygen in your compound?

4. Based on your solution to Question 3, what is the empirical formula of magnesium oxide?

8

From Oil of Wintergreen to Salicylic Acid

Objectives

- Examine the chemical reaction by which methyl salicylate (oil of wintergreen), a liquid, is converted to salicylic acid, a solid.

- Calculate the theoretical yield expected from the reaction.

- Determine the limiting reactant in the reaction.

- Calculate the percent yield of the reaction.

Introduction

Most chemical reactions are performed such that the amount of one reactant, typically the most expensive reactant, controls the amount of product that may be produced in a reaction. This reactant is known as the **limiting reactant**, or **limiting reagent**. The limiting reactant is the only reactant that is completely consumed during the reaction; all other reactants are in excess, and some will be left over once the reaction is finished. For example, consider the following balanced chemical equation:

$$A + B \longrightarrow C$$

Substances A and B react in a 1:1 molar ratio.

Suppose 1 mol of A and 1 mol of B are used in this reaction. Based on the stoichiometry of the balanced equation, we would expect that 1 mol of C would be produced. If, on the other hand, we started with 1 mol of A and only 0.5 mol of B, there would not be enough B to form 1 mol of C. Substance B limits the amount of C that can be produced and therefore is the limiting reactant. There would be 0.5 mol of A remaining when the reaction is complete, and only 0.5 mol of C would be produced.

As shown in this example, the amount of product formed in a reaction is controlled by the limiting reactant. Product stops forming when the limiting reactant is used up. The *calculated* amount of product that should be produced is called the **theoretical yield.** This is the maximum amount of product that can be produced from the quantities of reactants used. Typically the amount of product predicted is seldom obtained because of experimental error, side reactions (other reactions that consume some of the reactants or products), and possibly other causes. The *actual* amount of product obtained in a reaction is the **actual yield** and is always equal to or less than the theoretical yield. Often the theoretical yield and actual yield are compared and expressed as a percent called the **percent yield.**

$$\frac{\text{Actual yield}}{\text{Theoretical yield}} \times 100\% = \text{percent yield}$$

The reaction you will perform in this experiment can be represented by the following balanced chemical equation:

$$\text{Methyl salicylate} + \text{water} \longrightarrow \text{salicylic acid} + \text{methyl alcohol}$$

Methyl salicylate, well known for its characteristic wintergreen odor, is used as a flavoring and as an ingredient in various ointments. Salicylic acid, which is used in the preparation of aspirin, shares aspirin's ability to relieve pain.

When methyl salicylate is heated in water, salicylic acid and methyl alcohol form slowly. You can obtain a faster reaction if you replace the water with an aqueous solution of sodium hydroxide (NaOH). When the reaction is complete, the sodium hydroxide can be removed chemically by the addition of sulfuric

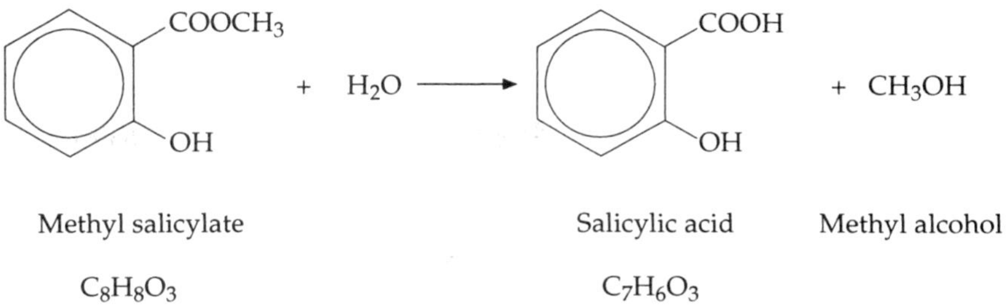

Figure 8.1
The reaction of methyl salicylate (oil of wintergreen)
with water to yield salicylic
acid and methyl alcohol.

Methyl salicylate

$C_8H_8O_3$

Salicylic acid

$C_7H_6O_3$

Methyl alcohol

acid (H_2SO_4). Although the conversion of methyl salicylate to salicylic acid
becomes slightly more complicated under these conditions, the overall results
are virtually identical with those shown in Figure 8.1.

The product that you have recovered may be impure and is often called the
crude product. You will use a technique called **recrystallization** to remove the
impurities. The crude solid product from the reaction is dissolved in a liquid
and crystallization is induced. Purification occurs if the impurities are more
soluble in the liquid than the desired compound is. The impurities will remain
in solution after the purified product has crystallized.

Experimental

Equipment: 250-mL beaker
10-mL graduated cylinder
100-mL graduated cylinder
marking pencil
balance
heating plate
50-mL Erlenmeyer flask
Büchner funnel
metal spatula
watch glass
filter paper for Büchner funnel

Materials: ice
methyl salicylate (oil of wintergreen)
6.0 M NaOH (sodium hydroxide)
8.0 M H_2SO_4 (sulfuric acid)

Safety Precautions

The solutions of sodium hydroxide and sulfuric acid must be handled carefully. These solutions can cause chemical burns in addition to ruining your
clothing. If you spill one of these solutions on you, wash the contaminated area

thoroughly and report the incident to your laboratory instructor. You may require further treatment.

Do not ingest any chemicals under any circumstances!

Procedure

Doing the Reaction

1. Transfer 60 mL of distilled water from a graduated cylinder to a 250-mL beaker. Mark the location of the upper surface of the water on the outer wall of the beaker, using a marking pencil. You will use this beaker for the reaction. First, however, discard the water and dry the beaker.

2. Obtain 4.0 mL of methyl salicylate in a clean, dry 10-mL graduated cylinder. Measure the mass using a balance and record it.

3. Add as much of the methyl salicylate from step 2 as possible to the dry, marked, 250-mL beaker from step 1.

4. Measure and record the mass of the 10-mL graduated cylinder. Calculate the mass of the methyl salicylate that you poured into the beaker.

5. Add 40 mL of 6.0 M NaOH from a clean 100-mL graduated cylinder. A white precipitate will form immediately. This precipitate is not the desired product. This precipitate is an ionic compound with the formula $NaOC_6H_4COOCH_3$.

6. Stir the mixture thoroughly with a clean glass stirring rod. Do not remove the stirring rod from the beaker.

7. Place the beaker on the heating plate, and heat the mixture to a *gentle* boil. Stir the solution occasionally. The precipitate will dissolve.

8. Continue to boil the solution for 15 min after the precipitate has dissolved. Use a stream of distilled water from a plastic wash bottle to rinse any solids from the inner walls of the beaker into the solution. However, do not let the total volume exceed 60 mL, as indicated by your mark from step 1.

9. Take this opportunity to assemble the glassware for suction filtration as discussed in the Introduction to this Manual.

10. After the period of heating is completed, remove the beaker *carefully* from the heating plate, and cool it in ice until it is only warm when you touch it. No precipitate should appear after this solution is cooled because the precipitate you started with is now gone and has reacted to form salicylic acid.

11. Add 50 mL of distilled water to the beaker; then *cautiously* add 50 mL of 8.0 M H_2SO_4 with stirring. A white precipitate of the crude product (salicylic acid) should form during the addition.

12. Cool the beaker again in ice until it is cold. At the same time, cool 50 mL of distilled water in an Erlenmeyer flask. You will use the cold water to wash the precipitate.

13. Filter the cold mixture by suction filtration, and wash the crude product with the cold water. The suction should be continued for several minutes to dry the precipitate partially. Wash and dry the 250-mL beaker during this time.

Recrystallizing the Product

1. Tilt the Büchner funnel over the 250-mL beaker. Use a metal spatula to separate carefully the precipitate from the filter paper and to transfer the precipitate to the beaker.
2. Add 100 mL of distilled water to the beaker from a graduated cylinder.
3. Place the beaker on the heating plate, and heat the mixture until a gentle boil occurs. Stir with a stirring rod. Several minutes of boiling are usually required before all of the precipitate dissolves.
4. Carefully remove the beaker from the heating plate, and allow the solution to cool for 5 min without disturbing it. You should see crystals of salicylic acid begin to appear. Record a description of their appearance.
5. Cool the beaker in ice until it is cold. At the same time, cool 50 mL of distilled water in an Erlenmeyer flask.
6. Filter the cold mixture by suction filtration. Wash the crystals with the cold water. Continue the suction for 15 min. Wash and dry the 250-mL beaker during this time.

Drying the Crystals

1. Measure and record the mass of the dry 250-mL beaker, using a balance.
2. Transfer the crystals to the beaker using the techniques that you used earlier.
3. Mark your name on the beaker with a marking pencil. Cover the beaker with a clean watch glass, and set it aside so the crystals can dry in the air until the next laboratory period.
4. When the crystals are dry, remove the watch glass. Measure and record the mass of the beaker and its contents.
5. Show your sample to your laboratory instructor. Obtain instructions for its disposal.

From Oil of Wintergreen to Salicylic Acid

NAME _____

SECTION _____

DATE _____

Prelaboratory Assignment

1. Draw the structures of methyl salicylate and salicylic acid, showing every chemical bond.

2. a. Calculate the number of moles of methyl salicylate molecules in 4.0 mL of methyl salicylate. Assume the density of methyl salicylate is 1.18 g/mL at 20°C. Show all calculations.

   b. What is the limiting reactant in this experiment? Why? Assume the two reactants are 40 mL of water and 4.0 mL of methyl salicylate.

3. What safety precautions must be observed during this experiment?

From Oil of Wintergreen to Salicylic Acid

NAME _____

SECTION _____

DATE _____

Results

Mass of methyl salicylate and graduated cylinder (g)　　_____

Mass of graduated cylinder (g)　　_____

Mass of methyl salicylate (g)　　_____

Description of crystals of salicylic acid

Mass of salicylic acid and beaker (g)　　_____

Mass of beaker (g)　　_____

Mass of salicylic acid (g)　　_____

Questions (Show all work!)

1. What is the actual yield of salicylic acid?

2. Calculate the theoretical yield of salicylic acid based on the *actual mass* of the reactants. Remember that you deduced the limiting reactant in the Prelaboratory Assignment.

3. Calculate the percent yield.

9
Spectrochemical Analysis of Iron Thiocyanate in Solution

Objectives

- Understand the principles of spectrochemical analysis.

- Using a spectrophotometer, determine how much light is absorbed by solutions containing different concentrations of iron thiocyanate.

- Determine the concentration of an unknown solution of iron thiocyanate.

Introduction

Some of the earliest means used for characterizing substances were based upon the observation of color. Color arises from the absorption and emission of light by matter, each form of matter displaying its own absorption and emission properties and, therefore, its own color. From this, it is a simple extension to realize that chemical species can be characterized by means of the color of their absorption or emission. In addition, the *amount* of light of a particular color absorbed by the sample is directly dependent on the concentration of the chemical species that absorbs that color. Typically, spectrochemical methods are much more quantitative than other chemical methods for analyzing trace quantities of chemical species.

The mathematical relationship that describes absorption of light is known as the Beer–Lambert law:

$$A = abc$$

where A = the unitless amount of light absorbed (**absorbance**) by the chemical species in the sample, a = absorptivity constant ($M^{-1}cm^{-1}$), b = the distance the light travels through the sample (cm), and c = the concentration of the chemical species in the sample (M).

White light is a combination of light of all colors. The Beer–Lambert law applies only at some specific wavelength that is a small fraction of the total spectral range. To take advantage of this law, you first will measure the absorbance of a set of solutions of known concentration and make a calibration curve by plotting absorbance versus concentration. Then the absorbance of a sample containing an unknown concentration of iron(III) thiocyanate will be measured and its concentration determined from the calibration curve.

Spectrophotometers typically have the same general design. A beam of light (radiation) from some source passes through a wavelength selector. The sample is placed between the wavelength selector and a light detector. The beam of light of some specified wavelength goes through the sample, and its intensity is measured at the detector. The spectrophotometer calculates the amount of light absorbed by the sample by measuring the difference between the intensity of the beam of light going into the sample and the intensity of the beam coming out of the sample.

Under usual conditions, iron(III), Fe^{3+}, and thiocyanate, SCN^-, are almost colorless, that is, they don't absorb light of any wavelength in the visible spectrum. You will react Fe^{3+} with thiocyanate, which will form a deeply colored product, iron(III) thiocyanate, $Fe(SCN)^{2+}$, which absorbs the maximum amount of light at a wavelength of 450 nm. The absorbance of $Fe(SCN)^{2+}$ can be directly related to its concentration.

Graphing

Much of the data you collect in this experiment will be presented in the form of a graph. Here are some pointers for making successful graphs, as previously discussed in Experiment 2:

1. Title the graph, and label all the axes with units (for example, $\mu g\ Pb^{2+}/mL$).
2. Spread the axes out so that the data covers the entire graph. You will need to decide on the size of divisions to mark on your graph.
3. Consider what these divisions will mean in terms of each box on the graph paper. A common student error is to make the divisions unequal.
4. Plot all points carefully and neatly. Use a pencil for graphing.
5. Do not "connect-the-dots"! If your graph represents a straight line (as for the calibration curve), use a ruler and draw the best straight line through your data points. This line does not have to intersect all the data points. Also, your line does not have to go through the origin (i.e., the zero–zero point). If your data are scattered, estimate where to draw your best straight line. Roughly an equal number of points should be above the line and below the line. This approximates a mathematical technique called *linear regression*, which judges where to draw the line to minimize the distance from each point to the line.
6. When you are reading data (for your unknown) from the graph, it is a good idea to use a ruler to read the data point by drawing a dotted line from the *x*-axis to the point. Do the same from the point to the *y*-axis.

A Word About Molarity, Volume, and Dilution

The unit of molarity (M) is mol/L. When you measure the volume of a solution with a known molarity you can calculate the number of moles because

$$mol/L \times L = mol$$

For example, 2.2 L of a 3.0 M solution of sodium hydroxide contains 6.6 moles of NaOH because

$$3.0\ mol/L \times 2.2\ L = 6.6\ mol$$

In this experiment, you will encounter dilution. Suppose you dilute a solution of known initial molarity (M_i) and known initial volume (V_i) to a solution with a final molarity (M_f) and final volume (V_f). The following equation is applicable:

$$M_i \times V_i = M_f \times V_f$$

In this experiment, you will know M_i, V_i, and V_f. You will be able to calculate M_f from this equation by rearranging it:

$$M_f = \frac{M_i \times V_i}{V_f}$$

Experimental

Equipment:	spectrophotometer (such as a Spectronic 20)
	5-mL graduated (Mohr) pipet
	pipet pump
	5 large (18 mm $\times$ 150 mm) test tubes with rubber stoppers
	6 small (13 mm $\times$ 100 mm) test tubes
	4 100-mL beakers
	100-mL graduated cylinder
	5 micropipets plus 1 bulb
Materials:	1 M KSCN) (potassium thiocyanate)
	0.0025 M Fe(NO$_3$)$_3$ [iron(III) nitrate] in 0.1 M HNO$_3$
	0.1 M HNO$_3$) (nitric acid)
	distilled water
	solution containing an unknown concentration of Fe(SCN)$^{2+}$ [iron(III) thiocyanate]

Safety Precautions

Nitric acid can cause chemical burns in addition to ruining your clothes. If you spill any nitric acid on you, wash the contaminated area thoroughly with tap water, and report the incident to your instructor. You may require further treatment.

Procedure

Getting Started

1. Obtain 5 large test tubes with rubber stoppers, a 5-mL graduated (Mohr) pipet, and a pipet pump. Also obtain 6 small test tubes.
2. Mark each large test tube with an identification number (1–5).
3. Obtain 30 mL of 1 M KSCN in a labeled beaker, 15 mL of 0.0025 M Fe(NO$_3$)$_3$ (in 0.1 M HNO$_3$) in another labeled beaker, and 32 mL of 0.1 M HNO$_3$ in a third labeled beaker.
4. Use the pipet to transfer 4.0 mL of the Fe(NO$_3$)$_3$ solution to a 100-mL graduated cylinder. Add distilled water until the meniscus is slightly below the 100-mL line on the cylinder. Use a micropipet to add enough water to make the bottom of the meniscus coincide with the 100-mL line. *Carefully* mix the solution in the graduated cylinder with a clean glass rod. Pour

the solution into a labeled beaker [the label can say diluted $Fe(NO_3)_3$] or Erlenmeyer flask, and mix thoroughly.

5. Use your pipet and the solution from step 4 to make the solutions listed in the following table. Use the directions in steps 5–11.

Test Tube No.	Diluted $Fe(NO_3)_3$ (mL)	1 M KSCN (mL)	0.1 M HNO$_3$ (mL)
1	1.0	5.0	4.0
2	2.0	5.0	3.0
3	3.0	5.0	2.0
4	4.0	5.0	1.0
5	5.0	5.0	0.0

6. Rinse your pipet with distilled water and then with several *small* portions of the diluted $Fe(NO_3)_3$ solution from step 4. Discard each rinse portion.
7. Add the volumes of the diluted $Fe(NO_3)_3$ solution shown in the table to the large test tubes using your pipet.
8. Rinse the pipet with distilled water and then with several *small* portions of the 1 M KSCN solution. Discard each rinse portion into the waste bottle.
9. Add the volumes of the 1 M KSCN solution shown in the table to the large test tubes using your pipet.
10. Rinse the pipet with distilled water and then with several *small* portions of the 0.1 M HNO$_3$ solution. Discard each rinse portion into the waste bottle.
11. Add the volumes of the 0.1 M HNO$_3$ solution shown in the table to the large test tubes using your pipet.
12. Insert the rubber stoppers in the test tubes and mix the solutions in each tube.

Obtaining the Calibration Curve from your Fe(SCN)$^{2+}$ Solutions

1. Label the small test tubes *at the top* with identification numbers 1–5 and U (for unknown).
2. Hold the small test tubes *only* at the very top, because fingerprints interfere with absorbance reading. As a precaution, wipe the outside of the tube with a Kimwipe. This will clean off the fingerprints.
3. Transfer the mixed solution from large test tube 1 to the small test tube 1 using a clean, *dry* micropipet. Fill the small test tube *at least* ¾ full.
4. Repeat step 3 with the rest of the solutions (2–5), i.e., transfer the mixed solutions from large test tubes into clean, dry, small test tubes.
5. Procedure for using a Spectronic 20 spectrophotometer: (IMPORTANT: The concentration mode of the Spectronic 20 does not give the concentration of your sample. Do not use a Spectronic 20 in the concentration mode!) Ask your instructor for instructions if you are using a spectrophotometer other than a Spectronic 20.

   a. Set the wavelength to 450 nm.

b. Set the mode to transmittance. (Nothing should be in the sample compartment, and the compartment should be closed.)

c. Adjust the zero control (left-hand knob) to a reading of 0.00.

d. Insert a small test tube filled with distilled water.

e. Adjust the transmittance control (right-hand knob) to a reading of 100.

f. Set the mode to absorbance.

g. Remove the test tube containing pure water, insert small test tube 1, and record the absorbance reading.

h. Repeat step 5g for samples 2–5.

6. Plot the absorbance against the calculated values of $[Fe(SCN)^{2+}]$ (see the Prelab problem 3 for example calculations) for all 5 test tubes. Draw the best possible straight line through the 5 points.

Obtaining the Concentration of an Unknown Solution

1. Obtain a sample of unknown from your laboratory instructor, record its number, and put the sample into the small test tube labeled U.

2. Measure the absorbance of your unknown as described in the previous section, step 5.

3. Using the corresponding absorbance for your unknown (test tube U) and your calibration graph, determine the $Fe(SCN)^{2+}$ concentration in your unknown. Record this on your report sheet.

9
Spectrochemical Analysis of Iron Thiocyanate in Solution

NAME _____

SECTION _____

DATE _____

Prelaboratory Assignment

For Problems 1–3 show work and include correct units!

1. Using the Beer–Lambert law, determine the concentration of chemical X in water, given that the amount of light absorbed by X is 0.855, the diameter of the test tube is 1 cm, and the absorptivity constant is $8.91 \times 10^3 \ M^{-1}cm^{-1}$.

2. Suppose a student made up a calibration sample by mixing 15.0 mL of $0.100 \ M$ $Fe(NO_3)_3$ with 10.0 mL of 10.0 M KSCN. What would be the concentration of $Fe(SCN)^{2+}$ in this sample? Notice that $Fe(NO_3)_3$ is the limiting reactant.

3. Calculate the final molarity of Fe^{3+} in test tubes 1, 2, and 3. Assume that the stock Fe^{3+} solution has a concentration of 1×10^{-4} M. In test tube 1, 1.0 mL of the stock Fe^{3+} solution is diluted to a final volume of 10.0 mL. In test tube 2, 2.0 mL of the stock Fe^{3+} solution is diluted to a final volume of 10.0 mL. In test tube 3, 3.0 mL of the stock Fe^{3+} solution is diluted to a final volume of 10.0 mL.

9
Spectrochemical Analysis
of Iron Thiocyanate
in Solution

NAME _____

SECTION _____

DATE _____

Results

Calibration curve

Test Tube no.	Absorbance	Concentration (include units)
1	_____	_____
2	_____	_____
3	_____	_____
4	_____	_____
5	_____	_____

Unknown

Unknown no.   _____

Absorbance   _____

Concentration   _____

Questions

1. What effect would there be on your experiment if you forgot to close the lid on your sample compartment when taking absorbance readings?

2. This experiment could also be run successfully using flat 7-Up instead of water as a solvent. Pepsi would not work. What's the difference?

3. Why is it important to take all of the absorbance readings at 450 nm, that is, at the wavelength where $Fe(SCN)^{2+}$ absorbs the maximum amount of light?

10
Solubilities Within a Chemical Family

Objectives

- Examine the trends in relative solubilities of some compounds of the alkaline earth metals.

- Compare the solubilities of compounds of lead, a Group IVA element, with similar compounds of alkaline earth metals.

Introduction

The elements in the periodic table are arranged in order of increasing atomic number such that elements with similar physical and chemical properties recur at regular intervals. This is known as the **periodic law**. Elements with the same number and kind of valence electrons appear in the same vertical column, or **group** (family). The alkaline earth metals (Group IIA elements) have an identical outer electron (valence) configuration, ns^2 (where n is the principal quantum number). For example, the valence electron configuration for magnesium is $3s^2$, whereas that for calcium is $4s^2$.

Elements within the same group form compounds that have the same general formula. Thus, all the alkaline earth metals form salts with nitrate, hydroxide, chloride, bromide, and iodide whose general formulas are $R(NO_3)_2$, $R(OH)_2$, RCl_2, RBr_2, and RI_2, respectively. You will use the following reaction to determine the qualitative solubilities of the alkaline earth hydroxides, chlorides, bromides, and iodides in water:

$$R(NO_3)_2(aq) + 2NaX(aq) \longrightarrow RX_2(aq \text{ or } s) + 2NaNO_3(aq)$$

where R is either a Group IIA element (Be, Mg, Ca, Sr, or Ba) or lead (Pb), and X is either OH, Cl, Br, or I. Sodium nitrate ($NaNO_3$) is very soluble in water. Therefore, in the preceding reaction, if any precipitate forms, it must be due to RX_2, which would mean that RX_2 is not very soluble in water.

In contrast, lead can form cations with either a +4 or +2 charge. For example, the formulas of the oxides that form with lead would be PbO_2 and PbO.

The alkaline earth metals and lead also form sulfates, carbonates, oxalates, and chromates with general formulas of RSO_4, RCO_3, RC_2O_4, and $RCrO_4$, respectively. These compounds will be prepared by the following reaction:

$$R(NO_3)_2(aq) + Na_2Y(aq) \longrightarrow RY(aq \text{ or } s) + 2NaNO_3(aq)$$

where Y represents either sulfate, carbonate, oxalate, or chromate. The solubilities of the RY compounds will be determined in exactly the same way as for the RX_2 compounds.

The periodic law does not imply that all the properties of the elements in a group are identical; rather, *trends* in properties within a period are usually observed. For example, atomic size increases going down a column of main-group elements, whereas ionization energy (the energy to remove an electron from an atom) decreases. Since you will examine only the qualitative aspects of the solubilities of compounds, you will not observe trends directly. Instead, you will have to infer their presence. The following example should make this process a little clearer.

Let us examine the solubilites of chromates. You use a chemical reaction to produce 1.0×10^{-4} mol of $MgCrO_4$, $CaCrO_4$, $SrCrO_4$, and $BaCrO_4$ in 1000 mL of water. Only $BaCrO_4$ forms a precipitate. This experiment would imply that, if a trend exists, the solubilites may decrease as the atomic number of the Group IIA metal increases. As you can see in Table 10.1, this is correct.

Table 10.1   Solubilities of Some Chromates (mol/1000 mL H_2O)

$MgCrO_4$	$CaCrO_4$	$SrCrO_4$	$BaCrO_4$
9.9	1.2	5.9×10^{-3}	1.1×10^{-5}

This example is not intended to portray a general trend. In some instances, solubilities increase as the atomic number of the Group IIA metal increases.

Experimental

Equipment: 5 small (13 mm × 100 mm) test tubes
5-mL or 10-mL graduated cylinder
marking pencil or tape

Materials: 0.1 M $Mg(NO_3)_2$ (magnesium nitrate)
0.1 M $Ca(NO_3)_2$ (calcium nitrate)
0.1 M $Sr(NO_3)_2$ (strontium nitrate)
0.1 M $Ba(NO_3)_2$ (barium nitrate)
0.1 M $Pb(NO_3)_2$ (lead nitrate)
1.0 M NaOH (sodium hydroxide)
1.0 M NaCl (sodium chloride)
0.1 M NaBr (sodium bromide)
0.1 M NaI (sodium iodide)
0.1 M Na_2SO_4 (sodium sulfate)
0.1 M Na_2CO_3 (sodium carbonate)
0.1 M $Na_2C_2O_4$ (sodium oxalate)

Safety Precautions

Wash your hands thoroughly after using solutions containing lead or oxalate because they are poisonous.

Never use your finger as a stopper!

Procedure

Getting Started

1. Obtain 5 small test tubes. Mark each with an identification number (1 through 5).
2. Using a 5-mL or 10-mL graduated cylinder, measure and pour 1 mL of distilled water into each test tube. Mark the height of the water in each test tube with a small piece of tape or a marking pencil. Add an additional 1 mL of distilled water to each test tube, and mark the new height of the water. Discard the water into a sink.

3. Obtain directions for discarding the solutions that you will use in this experiment.

Determining the Qualitative Solubilities

1. Using the lower marks on the test tubes as guides, add 1 mL of 0.1 M $Mg(NO_3)_2$ to test tube 1, 1 mL of 0.1 M $Ca(NO_3)_2$ to test tube 2, 1 mL of 0.1 M $Sr(NO_3)_2$ to test tube 3, 1 mL of 0.1 M $Ba(NO_3)_2$ to test tube 4, and 1 mL of 0.1 M $Pb(NO_3)_2$ to test tube 5.

2. Using the upper marks as guides, add 1 mL of 1.0 M NaOH to each test tube. Shake each tube gently and wait about 30 s. Note the colors of all precipitates. If any of the precipitates are virtually colorless, they may be difficult to see. Be observant! Note the relative amounts of the precipitates. If the solubility of a compound is barely exceeded, only a trace of a precipitate will appear. Record your observations.

   CAUTION: Do not use your finger as a stopper.

3. Wash the test tubes carefully and rinse them with distilled water.

4. Repeat steps 1–3 in turn with 1.0 M NaCl, 0.1 M NaBr, 0.1 M NaI, 0.1 M Na_2SO_4, 0.1 M Na_2CO_3, and 0.1 M $Na_2C_2O_4$ instead of the NaOH solution.

10
Solubilities Within a Chemical Family

NAME ____________________

SECTION ____________________

DATE ____________________

Prelaboratory Assignment

1. Give the names and symbols of the alkaline earth metals you will encounter in this experiment.

2. a. What is the general electron configuration of the alkaline earth metals?

   b. What is the electron configuration of lead?

   c. Compare the formulas of the oxides formed by the alkaline earth metals with those of lead.

3. Give general formulas for the following compounds with alkaline earth metals or lead.

a sulfate   an oxalate

a chloride   a hydroxide

a carbonate   a nitrate

an iodide   a chromate

a bromide

4. What general method will be used to examine the qualitative solubilities in this experiment?

5. What safety precautions must be observed in this experiment?

Results

	$Mg(NO_3)_2$	$Ca(NO_3)_2$	$Sr(NO_3)_2$	$Ba(NO_3)_2$	$Pb(NO_3)_2$
NaOH					
NaCl					
NaBr					
NaI					
Na_2SO_4					
Na_2CO_3					
$Na_2C_2O_4$					

Questions

1. What trends in the solubilities of the compounds of the alkaline earth metals can be inferred?

2. a. Compare the solubilities of the lead compounds with those of the alkaline earth metals. How are the solubilities similar, and how do they differ?

   b. As you have seen, lead can form the same kinds of compounds as the alkaline earth metals. Nevertheless, the solubilities may differ markedly. Give a convincing reason.

11
Chemical Bonding

Objectives

- Be able to draw Lewis (dot) structures for most simple molecules and ions.
- Be able to predict the three-dimensional molecular structure of simple molecules and polyatomic ions.

Introduction

A chemical **bond** is the force that holds groups of two or more atoms together and makes them function as a unit. Bonding involves only the valence (outermost) electrons of atoms. Valence electrons are transferred when a metal and a nonmetal react to form an ionic compound. Valence electrons are shared between atoms in covalent bonds.

The **Lewis structure** is a representation of a molecule that shows how the valence electrons are arranged among the atoms in a molecule. In writing Lewis structures, we include only the valence electrons, which are represented as dots. For example, hydrogen chloride, HCl, has the following Lewis structure:

$$H\!-\!\ddot{\underset{..}{Cl}}:$$

The line indicates 2 shared electrons.

The ionic compound NaCl has the following Lewis structure:

$$Na^+ \qquad [:\!\ddot{\underset{..}{Cl}}\!:]^-$$

No dots are associated with Na$^+$ because it has lost its only valence electron. The Cl$^-$ ion is shown with 8 electrons because it has a filled valence shell. No electrons are shared because NaCl is an ionic compound (electrons have been completely transferred).

The rules for writing Lewis structures are based on the observation that the most important requirement for the formation of a stable compound is that the atoms achieve a noble-gas electron configuration. Based on this observation, a few general rules will help you draw correct diagrams:

1. A hydrogen atom always forms one bond, that is, hydrogen forms a stable bond by sharing 2 electrons. It is said to follow the *duet rule.*
2. All other elements (except the transition metals) tend to acquire a noble-gas structure of 8 valence electrons by sharing electrons. They follow the *octet rule.*
3. Group 1A elements typically form one bond, and Group 2A elements typically form two bonds.
4. Group 4A and 5A elements typically form four bonds and three (sometimes four) bonds, respectively; for example, the carbon atom normally forms four bonds. When several carbon atoms appear in the same molecule, they are often bonded together.
5. Group 6A elements generally form two bonds, and Group 7A elements generally form one bond.
6. Noble gases do not form bonds because their valence orbitals are already filled.

With a little practice you will be able to draw Lewis structures for most molecules by inspection. For more complex molecules you may find the following procedure helpful:

1. Use one pair of electrons to form a bond between each pair of bound atoms. The preceding general rules may be helpful at this point.
2. Place electron dots around each element except hydrogen so the total number of electrons (bonding and nonbonding) for each atom is 8. Hydrogen requires only 2 electrons.
3. Count the number of electrons in your diagram.
4. Obtain the total number of valence electrons available by summing the valence electrons contributed by each atom in the molecule.
5. Compare the numbers from steps 3 and 4. If they are the same, the structure is complete. It they differ, the structure must be altered, typically by changing the number of multiple bonds.

Example Problems— Lewis Structure

Problem 1. Draw the Lewis structure for PH_3. (Follow the steps above.)

Step 1.   Knowing hydrogen forms only one bond, we draw a tentative diagram for the molecule, using lines to represent bonds. Remember that *each line represents 2 shared electrons.*

$$\begin{array}{c} H \\ | \\ H—P—H \end{array}$$

Step 2.   We place dots around phosphorus so that the total number of electrons is 8.

$$\begin{array}{c} H \\ | \\ H—P—H \\ \cdot\cdot \end{array}$$

Step 3.   3 bonds @ 2 electrons each = 6 electrons
Nonbonding electrons associated with P = 2 electrons
_____
          Total number of electrons available = 8 electrons

Step 4.   3 H atoms with 1 valence electron each = 3 electrons
                1 P atom with 5 valence electrons = 5 electrons
_____
          Total number of electrons available = 8 electrons

Step 5.   The number of electrons from steps 3 and 4 are equal, so the structure is complete.

$$\begin{array}{c} H \\ | \\ H\!-\!\!P\!-\!\!H \\ {\cdot\cdot} \end{array}$$

Problem 2. Draw the Lewis structure for CN^- (cyanide ion).

Step 1.   Carbon typically prefers 4 bonds: $[C\!\equiv\!N]^-$

Step 2.   Both C and N have associated with them 8 bonding electrons, so no electron dots need to be distributed.

Step 3.   4 bonds @ 2 electrons each = 8 electrons total

Step 4.

Carbon contributes 4 valence electrons =	4 electrons
Nitrogen contributes 5 valence electrons =	5 electrons
Single negative charge =	1 electron
Total number of electrons available =	10 electrons

Step 5.   The number of electrons from steps 3 and 4 do not agree. Change the number of multiple bonds from 4 to 3.

$$[:C\!\equiv\!N:]^-$$

This structure requires 10 electrons, and 10 electrons are available.

Valence Shell Electron Pair Repulsion (VSEPR) Model

Several models are used to describe the three-dimensional geometry of molecules and polyatomic ions. According to the VSEPR model, all pairs of electrons (bonding and nonbonding) are positioned as far apart as possible to minimize the electrostatic repulsion between electron pairs. The model describes the arrangement of every pair of electrons around the central atom.

When there are **two pairs** of electrons (bonding or nonbonding), the electron pair geometry is **linear**. When there are **three pairs** of electrons, the geometry is **trigonal planar**, and when there are **four pairs** of electrons, the geometry is **tetrahedral**.

Table 11.1 summarizes the possible geometries using the VSEPR model to describe how the **electron pairs are arranged** around the central atom.

Table 11.1
Arrangement of Electron Pairs in VSEPR Model

Number of Electron Pairs (Bonding and Non-bonding)	Electron Pair Geometry
2	Linear
3	Trigonal planar
4	Tetrahedral

Molecular Model   The molecular model describes the three-dimensional **arrangement of atoms** around the central atom. A systematic procedure to determine the molecular structure of a molecule or polyatomic ion follows.

Step 1.   Draw the Lewis structure.
Step 2.   Count the number of pairs of electrons that are bonding and the number that are nonbonding (lone pairs).
Step 3.   Determine the molecular model type and structure based on the number of bonding and nonbonding pairs of electrons around the central atom. See Table 11.2.

Table 11.2
Arrangement of Electron Pairs in Molecular Model

Number of Bonding Pairs of Electrons	Number of Nonbonding Pairs of Electrons	Molecular Model Type
4	0	Tetrahedral
3	0	Trigonal planar
3	1	Trigonal pyramid
2	0	Linear
2	1	Bent
2	2	Bent

Example Problem   What are the VSEPR and molecular models of the polyatomic ion, ClO_3^-? (Follow the steps above):

Step 1.   The Lewis structure (based on 26 available electrons):

$$: \overset{..}{\underset{}{O}} :$$
$$\overset{..}{:} O - Cl - \overset{..}{O} :$$

Step 2.   The number of nonbonding (lone) pairs of electrons surrounding Cl is 1, and the number of bonding pairs is 3.
Step 3.   The VSEPR model would be based on a total of 4 pairs of electrons surrounding the central Cl atom, so the arrangement would be tetrahedral. The molecular model would be trigonal pyramid.

Procedure

1. Work with a partner.
2. Obtain a molecular model kit and instructions for its use.
3. For each molecule and polyatomic ion listed in the Results section, draw the Lewis structure. Fill in the name of the three-dimensional structure based on both the VSEPR and molecular models. Use the molecular model kit to verify the geometry you predicted.

12
Boyle's Law and the Empty Space in Air

Objectives

- Examine the validity of Boyle's law, using air as the gas.
- Determine the amount of empty space in a sample of air.

Introduction

Unlike liquids and solids, gases are easily compressed. This property was first studied by Robert Boyle (1627–1691), who formulated the relationship that was later known as Boyle's law. When the pressure on a gas increases, the volume of the gas must decrease; when the pressure on a gas decreases, its volume must increase. In other words, Boyle's law states that at a constant temperature, the volume of a fixed quantity of gas is inversely proportional to the pressure. This law can be expressed mathematically as:

$$PV = k$$

where P is pressure, V is volume, and k is a constant. At a later date it was postulated that gases are easily compressed because most of their volume is empty space.

During this experiment you will measure the pressure and volume of a sample of air using a device called a *manometer*. This simple apparatus consists of two rigid glass tubes connected by a piece of flexible plastic tubing, as shown in Figure 12.1. The glass tubes contain mercury. One glass tube is open to the atmosphere; the other is equipped with a stopcock.

When the stopcock is open, both columns of mercury have identical heights, as in Figure 12.1A, since only atmospheric pressure is exerted on each column. Typically the stopcock is closed to trap a sample of air in one end of the manometer and to allow pressures less than or greater than atmospheric pressure to be applied to that sample of air.

Figure 12.1
A flexible-tube manometer. (A) $P = P$(atm), (B) $P > P$(atm), (C) $P < P$(atm), where P is the pressure exerted on the gas trapped in the manometer, and P(atm) is the atmospheric pressure.

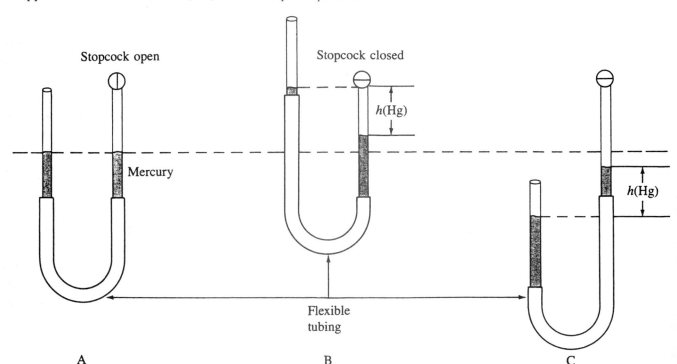

Stopcock open

Stopcock closed

h(Hg)

Mercury

h(Hg)

Flexible tubing

A   B   C

The flexible tubing allows the open-ended side of the manometer to be raised or lowered. Raising or lowering the open side affects the pressure exerted on the trapped air according to the relationship:

$$P = P(\text{atm}) + h(\text{Hg})$$

where P is the pressure exerted on the trapped air, $P(\text{atm})$ is the pressure of the atmosphere, and $h(\text{Hg})$ is the difference in the height of the mercury in the open and closed ends. A barometer is used to measure $P(\text{atm})$. To obtain the value of $h(\text{Hg})$, the height of the mercury in the closed end is subtracted from the height in the open end:

$$h(\text{Hg}) = h(\text{open}) - h(\text{closed})$$

When the open end is raised, as shown in Figure 12.1B, $h(\text{open}) > h(\text{closed})$, so $h(\text{Hg})$ is positive. Thus, the pressure exerted on the gas is greater than atmospheric pressure by $h(\text{Hg})$. When the open end is lowered, as shown in Figure 12.1C, $h(\text{closed}) > h(\text{open})$, and $h(\text{Hg})$ is negative. In this case, the pressure exerted on the trapped air is less than atmospheric pressure by $h(\text{Hg})$.

To measure the volume of the confined gas, we make the reasonable assumption that the cross-sectional area of the glass tube is constant throughout its length. The volume is then equal to the product of this area times the length of the trapped air column. Because the area is constant, the volume is proportional to the length of the air space. Therefore, the length is effectively a measure of the volume. Boyle's law can now be restated in terms of length instead of volume:

$$PL = c$$

where L is the length (in millimeters) of the gas column on the closed side, and c is a constant (equal to $k/\pi r^2$, or $k/$cross-sectional area of the glass tube).

In performing this experiment, you will first make a series of measurements of the length of a cylinder containing trapped air as you change the pressure by moving the open end of the manometer up and down. This experiment will provide the data for you to test Boyle's law. You should find that the product PL for each measurement is a constant within the precision of the measurement.

Second, you will determine how much empty space there is in another sample of air. The pressure exerted on this sample, and its volume and temperature, are obtained. By applying the ideal gas law, $PV = nRT$, where T is the absolute temperature in degrees Kelvin and R is the molar gas constant, to these experimental data, you can calculate the combined number of moles of N_2 and O_2 in the sample. From the calculated number of moles, Avogadro's number, and the volume occupied by a single molecule (assume a molecule of N_2 and a molecule of O_2 have identical volumes of 3.8×10^{-24} cm^3/molecule), you can compute the filled volume used by the molecules in your sample. Then you determine the empty space by subtracting the filled volume from the actual volume of the sample.

Experimental

Equipment: Boyle's law apparatus, also known as a manometer
               barometer
               thermometer
               100-mL graduated cylinder

Materials: none

Safety Precautions

Notify your instructor if any spills occur. Mercury spills must be cleaned up *by your instructor* immediately. Mercury vapor is very dangerous.

Procedure

Getting Started

1. Review the following definitions so that you understand the measurements that are required to test Boyle's law.

P	= pressure (mmHg) exerted on trapped air
$P(atm)$	= atmospheric pressure (mmHg)
$h(open)$	= height (mm) of mercury in the open side of the manometer
$h(closed)$	= height (mm) of mercury in the closed side of the manometer
$h(Hg)$	= $h(open) - h(closed)$
L	= length of gas column on the closed side, or, in other words, the result of subtracting $h(closed)$ from the measured height of the bottom of the stockcock

Testing Boyle's Law

1. Obtain and record the atmospheric pressure.
2. Obtain and record the height of the bottom of the stopcock.
3. Your laboratory instructor will close the stopcock and measure $h(open)$ and $h(closed)$ without moving the open end of the manometer. Record the results as Trial 1.
4. Your instructor will raise the open end to three successively higher positions. Record the measurements of $h(open)$ and $h(closed)$ at these positions as Trials 2, 3, and 4, respectively. Be sure to allow time for thermal equilibration between measurements.
5. Your instructor will now move the open end to three positions that are successively lower than the one recorded in step 3. Record the measurements of $h(open)$ and $h(closed)$ at these positions as Trials 5, 6, and 7, respectively. Again, be sure to allow a little time between measurements for thermal equilibrium.

6. Your laboratory instructor will allow you to move the open end to any position that you desire. Measure h(open) and h(closed) at this position, and record these measurements as Trial 8.

7. Calculate h(Hg), P, and L for each trial. Finally, calculate the product PL for each trial. Use scientific notation. Is PL a constant, i.e., essentially the same for all 8 trials?

Determining the Empty Space in Air

1. Measure and record the temperature of the laboratory. Convert that temperature to an absolute temperature on the Kelvin scale.

2. Place a 100-mL graduated cylinder on the laboratory bench in front of you. Think about the air that occupies 50.0 mL of the volume of the graduated cylinder. What is the pressure exerted on this volume of air? Record your result.

Boyle's Law and the Empty Space in Air

NAME _____

SECTION _____

DATE _____

Prelaboratory Assignment

1. a. What is Boyle's law in its usual form?

   b. What is the form of Boyle's law used in this experiment?

2. The atmospheric pressure on a particular day is 736 mmHg. Complete the following table for a gas trapped in the closed end of a manometer.

h(open) (mm)	h(closed) (mm)	h(Hg) (mmHg)	P (mmHg)
244	278	_____	_____
352	318	_____	_____

Calculations:

3. What special safety precaution must be observed during this experiment?

Boyle's Law and the Empty Space in Air

NAME _____

SECTION _____

DATE _____

Results

1. *Data for testing Boyle's law*

   Atmospheric pressure _____ mmHg

   Height of the bottom of the stopcock _____ mm

Trial	h(open) (mm)	h(closed) (mm)	h(Hg) (mmHg)	P (mmHg)	L (mm)	PL (mmHg · mm)
1	_____	_____	_____	_____	_____	_____
2	_____	_____	_____	_____	_____	_____
3	_____	_____	_____	_____	_____	_____
4	_____	_____	_____	_____	_____	_____
5	_____	_____	_____	_____	_____	_____
6	_____	_____	_____	_____	_____	_____
7	_____	_____	_____	_____	_____	_____
8	_____	_____	_____	_____	_____	_____

Calculations:

2. Data for determining the empty space in air

    Laboratory temperature   _____°C; _____K

    Pressure exerted on sample of air   _____ mmHg

    Volume of sample of air   _____ mL

Questions

1. To determine the relationship between two variables (in this experiment the two variables are pressure and volume), all other variables that might influence either of the two must be held constant. What are the two variables that are constant in your experiment?

2. Describe the relationship between pressure and volume shown by the results of your experiment. Be as quantitative as possible. Include the conditions that must be satisfied for your summary statement to be true (Question 1).

3. This question concerns the empty space in air.

   a. Calculate the number of moles of gas in your sample using the ideal gas law ($n = PV/RT$).

   b. Calculate the number of molecules of gas in your sample.

   c. Calculate the volume that is actually occupied by these molecules. (Assume that each molecule occupies 3.8×10^{-24} cm^3.)

   d. Calculate the empty space in your sample.

13
Solids and Liquids

Objectives

- To determine the freezing point of a compound utilizing a cooling curve.
- To determine the boiling points of three related compounds and observe any trends.

Introduction

What happens to a liquid when it is cooled? At a certain temperature, the **freezing point**, the liquid solidifies. For example, water becomes ice at 0°C. When water is cooled, its temperature decreases until it reaches 0°C, when the liquid begins to freeze. An interesting thing happens at the freezing point: even though cooling continues, the water temperature stays at 0°C until all of the water has turned to ice. Only when all of the water has changed to the solid state does the temperature begin to drop again. (You are then cooling ice.)

Figure 13.1 shows a typical **cooling/heating curve**. Going from left to right means energy is being removed, that is, the liquid is being cooled. In the case of water being cooled, t_f, the freezing point temperature, is 0°C.

Figure 13.1
Cooling curve demonstrating change of state.

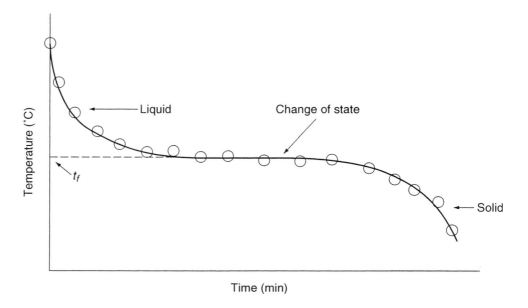

Water and ice can coexist indefinitely if the temperature is held at 0°C. Two physical states (liquid and solid) are present simultaneously. Theoretically this temperature, the freezing point, should have the same value as the melting point. The **melting point** is the temperature at which a solid becomes a liquid. Similarly, the **boiling point** (the temperature at which a liquid becomes a gas) should have the same value as the **condensation point** (the temperature at which a gas becomes a liquid). For water, the boiling point and condensation point occur at 100°C (assuming standard atmospheric pressure). Again, two physical states are present, water and steam.

In the first part of this experiment, liquid *para*-dichlorobenzene will be allowed to cool to a solid while the temperature/time relationship is recorded. The data will be plotted as a cooling curve. The temperature should remain constant as the liquid solidifies.

In the second procedure the boiling points of three related compounds will be determined. The compounds, methanol, ethanol, and 2-propanol (iso-propanol), are different types of alcohols; their structures are illustrated in Figure 13.2.

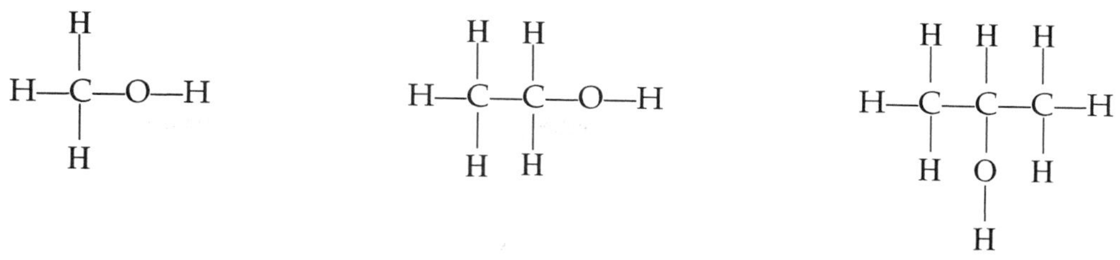

Figure 13.2
The structures of
methanol, ethanol,
and 2-propanol.

$$H-\overset{\overset{\displaystyle H}{|}}{\underset{\underset{\displaystyle H}{|}}{C}}-O-H$$

Methanol

$$H-\overset{\overset{\displaystyle H}{|}}{\underset{\underset{\displaystyle H}{|}}{C}}-\overset{\overset{\displaystyle H}{|}}{\underset{\underset{\displaystyle H}{|}}{C}}-O-H$$

Ethanol

$$H-\overset{\overset{\displaystyle H}{|}}{\underset{\underset{\displaystyle H}{|}}{C}}-\overset{\overset{\displaystyle H}{|}}{\underset{\underset{\displaystyle O}{|}}{C}}-\overset{\overset{\displaystyle H}{|}}{\underset{\underset{\displaystyle H}{|}}{C}}-H$$
$$\underset{\displaystyle H}{}$$

2-Propanol

Experimental

Equipment: 2 400-mL beakers
110°C thermometer
2 large (25 mm × 150 mm) test tubes
ring stand with clamps
wire gauze
hot plate
split rubber or cork stopper
boiling chips
aluminum foil

Materials: *para*-dichlorobenzene (*p*-dichlorobenzene)
methanol
ethanol
2-propanol (isopropanol)

Safety Precautions

Do not force the thermometer into a bored rubber stopper. Instead, use a split rubber stopper. Avoid contact with the *p*-dichlorobenzene. Wash hands thoroughly after completing work with *p*-dichlorobenzene. Do not use a Bunsen burner to heat any of the alcohols; they are extremely flammable. *Do not consume* any of the materials used in this experiment. Some alcohols are **deadly** poisons! Be careful not to hit the thermometer against hard surfaces, as the bulb can easily break.

Procedure

Getting Started

1.   Add 300 mL of distilled water to a 400-mL beaker. Heat this water, using a heating plate, to 40°C, and then remove the beaker from the heating plate.
2.   Setup the apparatus as shown in Figure 13.3, using the heated water from step 1.
3.   Obtain a test tube containing *para*-dichlorobenzene.

Figure 13.3
Apparatus suitable
for determining the
freezing point of a
solvent or a solution.

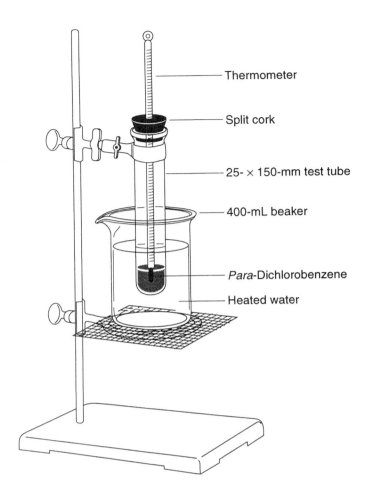

Thermometer

Split cork

25- × 150-mm test tube

400-mL beaker

Para-Dichlorobenzene

Heated water

**Freezing Point
Measurement
of *para*-
Dichlorobenzene**

1. In another 400-mL beaker, heat 300 mL of distilled water to a gentle boil using the heating plate. Melt the *para*-dichlorobenzene by immersing the bottom of the test tube in the boiling water. Grip the test tube with a test tube holder.

2. Insert a thermometer into the melted *para*-dichlorobenzene, and continue heating until the temperature is well above 65°C. Do not insert the split stopper into the top of the test tube. The top of the test tube must be vented to the atmosphere.

3. Rapidly transfer the test tube with thermometer into the 400-mL beaker of heated water. Clamp the test tube so that its bottom is immersed in the water bath. Arrange the height of the test tube so that the level of the *para*-dichlorobenzene is close to that of the water in the beaker.

4. Suspend the thermometer into the test tube until the immersion line is at the liquid surface.

5. Begin thermometer readings when the temperature drops to 65.0°C. Record the temperature and the state of the compound every 30 s. To ensure a correct temperature reading, it is advisable to stir the compound as it cools. Before reading the temperature, move the thermometer up and down to mix the compound.

Figure 13.4
Apparatus suitable for determining the boiling point of a solvent.

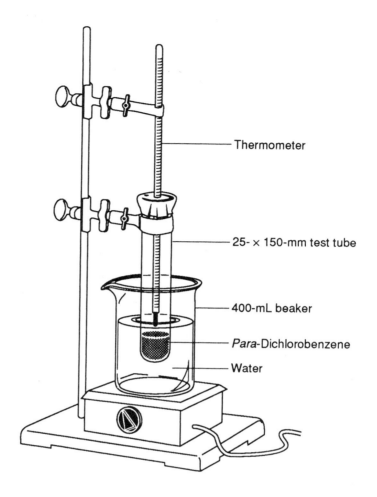

— Thermometer

— 25- × 150-mm test tube

— 400-mL beaker

— *Para*-Dichlorobenzene

— Water

6. Continue recording the temperature until the compound has solidified. Do not attempt to pull out the thermometer because it may break. Melt the *para*-dichlorobenzene until the thermometer is released. Do not pour out the liquid *para*-dichlorobenzene. Return it to your instructor.

Boiling Points of Methanol, Ethanol, and 2-Propanol

1. Fill a 400-mL beaker with 200 mL of water, and assemble the apparatus as shown in Figure 13.4.
2. Obtain a large, clean, dry test tube. Fill it halfway with methanol. Add two small boiling chips. Clamp the test tube so that its bottom is immersed in the water bath. Arrange the height of the test tube so that the surface of the methanol is below that of the water.
3. Carefully and securely clamp the thermometer with a thermometer clamp, and lower it into the test tube so that the bulb of the thermometer is about 1 cm above the surface of the methanol.
4. Cover the mouth of the test tube with a piece of aluminum foil to avoid evaporation of the liquid into the room.

CAUTION: Most of the organic liquids are flammable and toxic.

5. Heat the water bath slowly using the hot plate.
6. The methanol should start boiling. The vapor will rise to the tip of the thermometer, and you will notice a rise in temperature. Continue heating until the temperature no longer rises and remains constant. This is the boiling point of methanol. Record the temperature to the nearest 0.1°C.
7. Dispose of the methanol in the proper waste receptacle, and clean and dry the large test tube.
8. Repeat steps 2–7 using ethanol and then 2-propanol.

NAME _____

SECTION _____

DATE _____

Prelaboratory Assignment

1. What is the difference between the boiling point of water and the condensation point of water?

2. The freezing point of *para*-dichlorobenzene corresponds to which point on the cooling curve in Figure 13.1?

3. The graph depicts a cooling curve for cyclohexane. From the graph, estimate the melting point to the nearest 0.1°C.

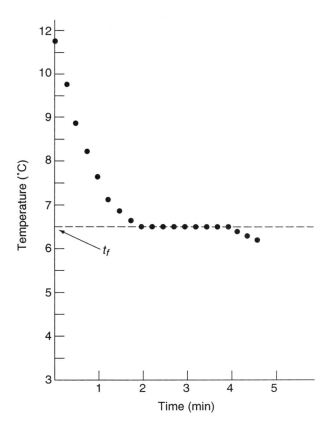

The melting point for cyclohexane is _____

4. What are the molecular weights of the three alcohols used in this experiment: methanol, ethanol, and 2-propanol?

5. What special safety precautions are cited in this experiment?

Results

Freezing point measurement of *para*-Dichlorobenzene

Temperature	Time (s)	State
65.0°C	0:00	liquid
	0:30	
	1:00	
	1:30	
	2:00	
	2:30	
	3:00	
	3:30	
	4:00	
	4:30	
	5:00	
	5:30	
	6:00	
	6:30	
	7:00	
	7:30	
	8:00	
	8:30	
	9:00	
	9:30	
	10:00	

Boiling points

Boiling point of methanol (CH$_3$OH)   _____

Boiling point of ethanol (C$_2$H$_5$OH)   _____

Boiling point of 2-propanol (C$_3$H$_7$OH)   _____

Questions

1. a. What is the freezing point of *para*-dichlorobenzene?

    b. What is the melting point of *para*-dichlorobenzene?

2. What trend do you observe in the boiling points of the three alcohols you studied?

14A
Softening Hard Water

Objectives

- Examine some reactions involving aqueous solutions containing Ca^{2+} and Mg^{2+} ions.

- Soften hard water by ion exchange.

Introduction

Our water comes from many sources such as lakes, rivers, springs, and wells. Although it may be fit to drink, water is almost never chemically pure. Typically drinking water contains a variety of dissolved ions including Ca^{2+}, Mg^{2+}, Fe^{3+}, HCO_3^-, and SO_4^{2-}. If the concentration of these ions is relatively large, that is, exceeding 150 parts per million (ppm), the water is called **hard water** (150 ppm means that there are 150 grams of these ions for every million grams of the total solution). Water with relatively small concentrations of these ions is referred to as **soft water**.

Calcium and magnesium ions are often found in minerals in the form of carbonates, $CaCO_3$ and $MgCO_3$. For example, the mineral limestone is mostly calcium carbonate, $CaCO_3$. Although these substances are not soluble in pure water, they are soluble in dilute solutions of carbonic acid (H_2CO_3) that occur in natural water. Atmospheric carbon dioxide (CO_2) dissolves in natural water to form this acid according to the equation:

$$CO_2(g) + H_2O(l) \longrightarrow H_2CO_3(aq)$$

The carbonic acid formed in this process reacts with calcium or magnesium carbonate to produce a soluble substance. For example, carbonic acid reacts with $CaCO_3$ to form soluble calcium hydrogen carbonate:

$$CaCO_3(s) + H_2CO_3(aq) \longrightarrow Ca^{2+}(aq) + 2HCO_3^-(aq)$$

A similar equation can be written for $MgCO_3$.

Hardness in water is undesirable for two reasons:

1. Soap does not form suds in hard water, as indicated in the following chemical equation:

$$Ca^{2+}(aq) \; + \; 2C_{17}H_{35}COO^-Na^+(aq) \; \longrightarrow \; Ca(C_{17}H_{35}COO)_2(s) \; + \; 2Na^+(aq)$$
$$\text{soap} \qquad\qquad\qquad\qquad \text{soap scum}$$

A similar equation can be written for Mg^{2+}. The dissolved cations (Ca^{2+} or Mg^{2+}) form insoluble, gray, curdy scum (bathtub ring) with soap; therefore soaps lose their cleansing power.

2. The reactions responsible for the solubility of $CaCO_3$ and $MgCO_3$ in dilute solutions of H_2CO_3 are readily reversed when heat is applied:

$$Ca^{2+}(aq) + 2HCO_3^-(aq) + heat \longrightarrow CaCO_3(s) + H_2O(l) + CO_2(g)$$

A similar equation can be written for Mg^{2+}. These solid carbonates that form are partly responsible for the residues that coat the inside surfaces of tea kettles, hot-water pipes, boilers, and heat exchangers.

A cation-exchange resin will be used in this experiment to make the hard water soft. This resin is an insoluble organic substance whose molecules consist

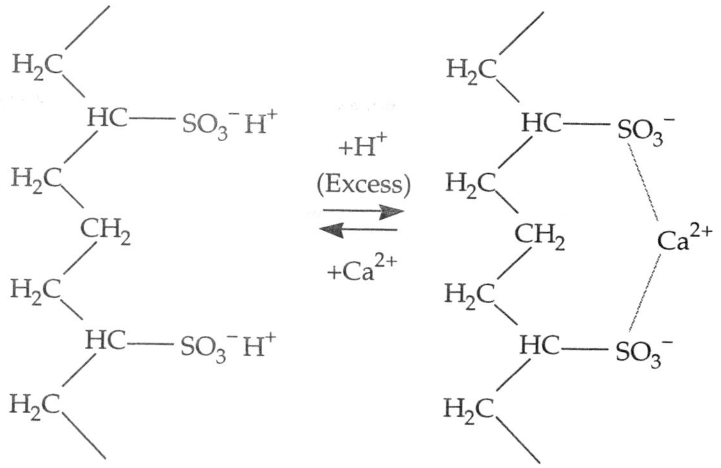

Figure 14A.1
A cation-exchange resin in the H+ form (left), Ca2+ form (right).

of long chains of atoms. The sulfonic acid group, $-SO_3H$, is chemically bonded to the chain in many places. Figure 14A.1 shows the mechanism for cation exchange where one type of ion simply replaces an another type of ion.

Ions with a +2 charge, such as Ca^{2+} ions, will be more strongly bound to the resin than will ions with a +1 charge, such as Na^+ or H^+ ions. As a result, a cation-exchange resin in the H^+ form will easily exchange ions with a dilute solution of Ca^{2+} and Mg^{2+} ions (like hard water). After this exchange has occurred, the resin in its original H^+ form can be regenerated, but a large excess of a strong acid must be used.

The hardness of the water will be measured by a technique called **titration**. Typically, titration involves the controlled addition of one solution from a buret to another compound with which the solution will react. In this experiment you will use a medicine dropper instead of a buret. An **indicator**, Eriochrome Black T, is used to signal when the titration arrives at the point at which the reactants are stoichiometrically equal, as defined by the reaction equation. Indicators are typically plant dyes that change color during a reaction. For example, in this experiment you will measure exactly how much dihydrogen ethylenediaminetetraacetic (H_2EDTA^{2-}) anion solution is required to convert completely all of the Ca^{2+} to $Ca(EDTA)^{2-}$:

$$H_2EDTA^{2-}(aq) + Ca^{2+}(aq) \longrightarrow Ca(EDTA)^{2-}(aq) + 2H^+(aq)$$

You will know when you have added enough H_2EDTA^{2-} because the indicator, Eriochrome Black T, in the reaction solution will turn from rose-pink to blue. In other words, the "harder" the solution, the higher the concentration of Ca^{2+} or Mg^{2+} in solution, and, therefore, the more H_2EDTA^{2-} required for the titration.

Since $CaCO_3$ and $MgCO_3$, along with CO_2, play important roles in the formation of hard water and in some of its effects, the first part of the experiment will focus on these substances. The second part will involve the softening of water by ion exchange. To judge the hardness of the water sample before and after ion exchange, you will titrate the solutions with a solution containing H_2EDTA^{2-} and the indicator, Eriochrome Black T.

Experimental

Equipment: 4 small (13 mm × 100 mm) test tubes
marking pencil
blue litmus paper
250-mL beaker
50-mL Erlenmeyer flask
100-mL graduated cylinder
medicine dropper

Materials: 0.1 M Mg(NO$_3$)$_2$ (magnesium nitrate)
0.1 M Ca(NO$_3$)$_2$ (calcium nitrate)
0.1 M Na$_2$CO$_3$ (sodium carbonate)
0.01 M Na$_2$H$_2$EDTA (sodium ethylenediaminetetraacetate)
6.0 M HCl (hydrochloric acid)
NH$_3$–NH$_4$Cl (ammonia–ammonium chloride) buffer
Eriochrome Black T and NaCl
dry ice (solid CO$_2$)
artificial hard water (if needed)
cation-exchange resin (with –SO$_3$H groups)

Safety Precautions

Handle the solution of hydrochloric acid carefully. It can cause chemical burns in addition to ruining your clothing. If you spill any on you, wash the contaminated area thoroughly and report the incident to your laboratory instructor. You may require further treatment.

Procedure

Getting Started

1. Obtain 4 small test tubes and 2 pieces of blue litmus paper. Additional litmus paper can be obtained if you need it. Label your test tubes 1 through 4.
2. Obtain about 6 g of dry cation-exchange resin or about 9 g of wet resin in a 250-mL beaker. Since this resin is very expensive, it must be returned when you finish the experiment.
3. Your laboratory instructor will tell you whether to use tap water or artificial hard water for the ion exchange.

Observing Reactions of Ca²⁺ and Mg²⁺ Ions

1. Add 10 drops of 0.1 M Ca(NO$_3$)$_2$ to test tubes numbered 1 and 2. Add 10 drops of 0.1 M Mg(NO$_3$)$_2$ to test tubes labeled 3 and 4.
2. Add 10 drops of 0.1 M Na$_2$CO$_3$ to each test tube and shake gently. Observe the results carefully and record them. Be sure to note any differences between the results from test tubes 1 and 3.

3. Add dry ice (solid CO_2) slowly to test tubes 1 and 3. Use a quantity about the size of 2 peas. Shake each test tube gently until no further effects can be seen. Record the results.

4. Add 2 drops of 6.0 M HCl to test tubes 2 and 4. Record the results.

Preparing the Resin

1. Add 6.0 M HCl to the beaker containing the resin until the resin is barely covered. Allow the mixture to stand for about 2 min.

2. Add about 200 mL of distilled water (*not* tap water), and let the mixture stand until the resin settles to the bottom of the beaker.

3. Carefully decant (pour off) the solution above the resin until only a few milliliters of solution remain.

4. Wash the resin by adding 200 mL of distilled water. Let the resin settle to the bottom and then decant again.

5. Repeat step 4 three more times.

6. Repeat step 4 again, but test the solution with blue litmus paper before decanting. If the paper turns pink, continue washing according to step 4 until no change occurs when you test the solution with blue litmus paper.

Softening Hard Water

1. Cover the resin in the beaker with a quantity of tap water or artificial hard water. Use 100 mL unless your laboratory instructor provides other directions. If you are using artificial hard water, you will need to have enough for water softening (100 mL) plus an additional 50 mL.

2. Let the mixture of resin and water stand for 10 min with occasional stirring.

3. During this time, add 20 mL of unsoftened water (either tap water or artificial hard water) to a clean 50-mL Erlenmeyer flask from a clean graduated cylinder.

4. Add 5 mL of a NH_3–NH_4Cl buffer ($[OH^-] \approx 10^{-4}\ M$) to this flask.

5. Add a very small portion of a solid mixture of Eriochrome Black T and NaCl. This portion should not exceed $\frac{1}{8}$ in. on the tip of a metal spatula. The solution should turn light rose-pink color after this addition.

6. Add 0.01 M Na_2H_2EDTA *slowly* by drops from a medicine dropper, counting the drops and swirling the solution in the flask. As the end point is approached, you should see a lavender color. After this point, add the solution more slowly until the first appearance of the blue color. Record the number of drops that were required.

7. After 10 min have elapsed, carefully decant the softened water from step 2 into a clean beaker until only a few milliliters of the water remains with the resin.

8. Repeat steps 3–5 using the softened water from step 7 instead of unsoftened water. The blue color will appear immediately if virtually all of the Ca^{2+} and Mg^{2+} ions in the water have been removed. If so, record this result. If not, repeat step 6.

Finishing the Experiment

1. Convert the resin to the H^+ form, using the same method that you used to prepare the resin.

2. Return the resin to your laboratory instructor.

14A
Softening Hard Water

NAME ____________________

SECTION ____________________

DATE ____________________

Prelaboratory Assignment

1. Define the following terms:
   a. Hard water

   b. Soft water

   c. Water softening

   d. Ion exchange

   e. Cation-exchange resin

2. a. Give two detrimental effects that result from using hard water.

b. Write the net ionic equation that describes the formation of the insoluble curdy precipitate when soap is used in hard water.

3. Write the chemical equation that describes the titration of Mg^{2+} ions with H_2EDTA^{2-} ions in the presence of the indicator.

4. What will you do with the cation-exchange resin after you finish the experiment?

5. What safety precautions must be observed in this experiment?

14A Softening Hard Water

NAME ____________________

SECTION ____________________

DATE ____________________

Results

1. *Reactions of Ca^{2+} and Mg^{2+} ions*

	$Ca(NO_3)_2$	$Mg(NO_3)_2$
Na_2CO_3		
$Na_2CO_3 + CO_2$		
$Na_2CO_3 + HCl$		

2. *Softening hard water*

Number of drops of 0.01 *M* Na_2H_2EDTA:

Before ion exchange ____________

After ion exchange ____________

Questions

1. Write the complete ionic equation(s) for any precipitation reactions that you observed with Ca^{2+} and Mg^{2+} ions in the first part of the experiment.

2. Using the following equation, determine the percent efficiency of your water softening from the number of drops of Na_2H_2EDTA solution required before and after you treated the water with cation-exchange resin. Show all work.

$$\% \text{ efficiency} = \frac{(\text{drops before exchange} - \text{drops after exchange})}{\text{drops before ion exchange}} \times 100\%$$

14B

Analysis of Calcium in Milk

Objective

- Determine the amount of calcium in a sample of milk.

Introduction

Calcium, a chemical element and our body's #1 macromineral, has recently been very newsworthy. When we do not consume enough calcium, Ca^{2+}, our body takes the calcium it needs from our skeleton and uses it in our nervous, digestive, and circulatory systems. This is especially serious for women after menopause. People with this disorder, known as *osteoporosis*, often experience spinal fractures, significant height loss, pain, a stooped, humpbacked appearance, and an increased number of other fractures.

Calcium in the diet serves to slow down the rate of bone loss. The exact minimum amounts needed vary with age, sex, environment, heredity, and many other factors. The RDA (*Recommended Daily Allowance*) is only 800 mg Ca^{2+} per day. Not all calcium consumed is absorbed. Absorption may be blocked by the consumption of certain items such as coffee and alcohol. Absorption of calcium by the body is also related to the presence of vitamins C and D and the minerals magnesium and phosphorus. For women, the recommended amounts of Ca^{2+} are shown in Table 14B.1.

Table 14B.1
RDA of Calcium for Women

Age 11–18 yr	1000 mg
Premenopausal	1000 mg
Postmenopausal	
not on estrogen replacement therapy	1500 mg
on estrogen replacement therapy	1000 mg

Recent research has indicated that the form in which calcium is consumed also affects the body's absorption rate. Some studies show that the calcium compound added to some brands of orange juice is absorbed better than from other sources.

Leading sources of dietary calcium include dairy products, fish, and green leafy vegetables. Table 14B.2 lists the amounts of calcium in some foods.

In this experiment the calcium ion will react in a 1:1 mole ratio with EDTA (ethylenediaminetetraacetic acid). The balanced equation for this reaction is as follows:

$$H_2EDTA^{2-}(aq) + Ca^{2+}(aq) \longrightarrow Ca(EDTA)^{2-}(aq) + 2H^+(aq)$$

The final product, $[Ca(EDTA)]^{2-}$, is a complex ion; its structure is shown in Figure 14B.1. The human body cannot metabolize this form of calcium.

The type of analysis you will carry out is known as a **titration**. A titration is the controlled addition of one compound from a buret to another compound with which it will react. A known volume (10.0 mL) of milk will be titrated with a known molarity (0.0200 M) of EDTA solution delivered from a buret (a long tube with graduated markings designed to deliver fluid in a controlled fashion).

Table 14B.2

Calcium Content of Certain Foods

Food	Amount	Calcium (mg)
Low-fat milk, 1% or 2%	1 cup	310
Skim milk	1 cup	300
Whole milk	1 cup	290
Buttermilk	1 cup	290
Nonfat dry milk	2 tablespoons	105
Eggnog	1 cup	330
Plain yogurt (whole milk)	1 cup	300
Plain yogurt (low-fat)	1 cup	400
Mozzarella cheese, part skim	1 oz	210
Parmesan cheese	1 oz	340
Swiss cheese	1 oz	270
Cottage cheese, 2% fat	1 cup	160
Collard greens	1 cup	360
Turnip greens	1 cup	250
Kale	1 cup	200
Bok choy	1 cup	250
Broccoli	1 cup	150
Canned sardines (with bones)	4 oz	500
Canned red salmon	4 oz	290
Canned mackerel	4 oz	300

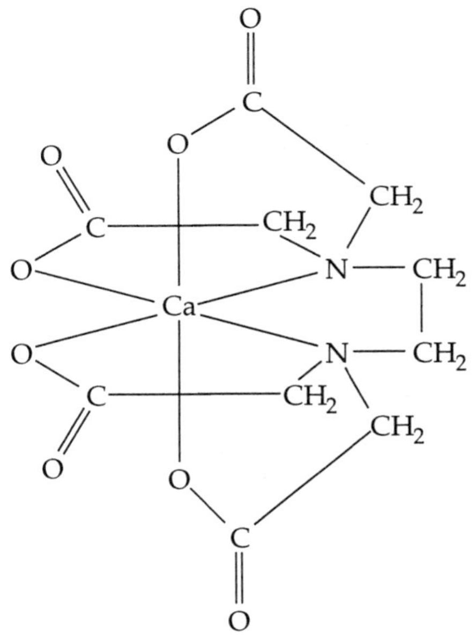

Figure 14B.1
The structure of the complex ion, $[Ca(EDTA)]^{2-}$, calcium ethylenediamine-tetraacetate.

The titration will be stopped when the **end point** is reached, signaling that all the calcium has reacted. An **indicator** is a chemical compound, often a plant dye, that changes color during a reaction. The indicator you will use, Eriochrome Black T, is a solid. The reaction will be done in a solution with a constant concentration of H^+, 1.0×10^{-10} M, which is a basic solution. The solution will also contain some $Mg(EDTA)^{2-}$, which will not affect the calculations but does make the end point more visible and easier to determine.

Various milk samples will be available. You will be given a sample to analyze for its calcium content. Afterwards you and your classmates will compare results to determine which sample had the most calcium.

Example Calculations

A student titrates a 10.0-mL milk sample with 0.0200 M EDTA solution and finds that it requires 17.0 mL of EDTA (Na_2H_2EDTA) to reach the end point.

1. How many milligrams of calcium are in the 10.0-mL sample?

$$17.0 \text{ mL EDTA} \times \frac{1 \text{ L EDTA}}{1000 \text{ mL}} \times \frac{0.0200 \text{ mol EDTA}}{1 \text{ L EDTA}} \times \frac{1 \text{ mol Ca}^{2+}}{1 \text{ mol EDTA}}$$

$$\times \frac{40.0 \text{ g Ca}^{2+}}{1 \text{ mol Ca}^{2+}} \times \frac{1000 \text{ mg Ca}^{2+}}{1 \text{ g Ca}^{2+}} = 13.6 \text{ mg Ca}^{2+}$$

2. How many milligrams of calcium are in a 8.00-oz serving of this milk?

$$8.00 \text{ oz} \times \frac{29.6 \text{ mL}}{1 \text{ oz}} \times \frac{13.6 \text{ mg Ca}^{2+}}{10.0 \text{ mL}} = 322 \text{ mg Ca}^{2+}$$

Experimental

Equipment: buret
              buret brush
              small funnel
              3 250-mL Erlenmeyer flasks
              100-mL or 250-mL beaker
              10.0-mL pipet with suction bulb or other device
              spatula
              marking pencil
              100-mL graduated cylinder

Materials: milk sample
              distilled water
              0.0200 M EDTA (Na_2H_2EDTA)
              Eriochrome Black T
              buffer solution

Procedure

Filling the Buret

1. Obtain approximately 100 mL of EDTA solution in a clean, dry beaker.
2. Obtain a buret, and check to see if it is clean by rinsing it with distilled water. If it is not clean (that is, if water droplets stick to the inside of the buret), wash the buret with soap, water, and a buret brush. Then rinse with distilled water, and allow the water to drain through the stopcock to rinse the stopcock.
3. Set up the buret as shown in Figure 14B.2.
4. Using a small funnel, pour 5 mL of EDTA solution into the buret. Tilt the buret to rinse all of the sides, and drain the liquid through the stopcock. Repeat this step two more times. If you do not adequately rinse the buret with EDTA solution, the EDTA in the buret will be diluted and more will be required for the titration, giving a large and incorrect value of calcium.
5. Fill the buret to above the zero mark with EDTA solution.
6. Open the stopcock wide for a few seconds to remove all air from the tip. If air still remains in the tip, ask your instructor for assistance. (If the air comes out during the titration, you will end up with a volume of EDTA that is too large and incorrect.)

Figure 14B.2
The correct arrangement for filling and using a buret held in place with a buret clamp attached to a ring stand.

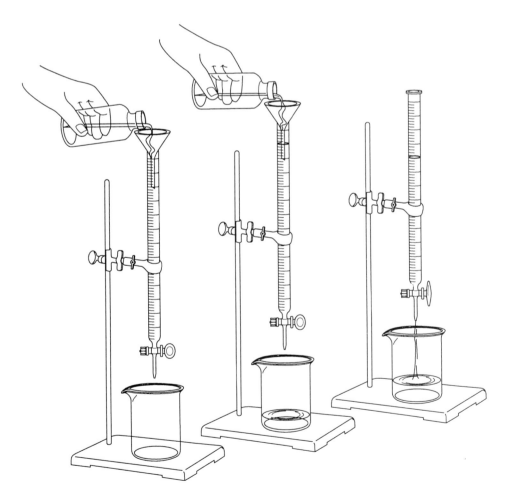

Figure 14B.3
The proper method of reading a meniscus to avoid parallax error.

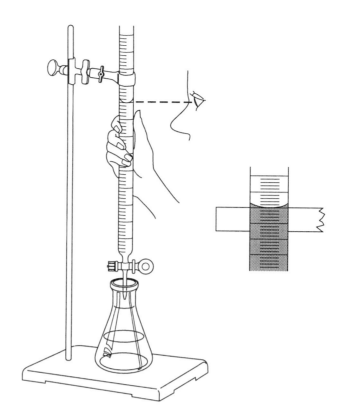

7. Refill the buret with more EDTA until the level is between the 0.00-mL and 1.00-mL markings. Do not attempt to set the level at exactly 0.00 mL!

8. Locate the bottom of the meniscus, and hold a darkened card in back of the buret. Have your eye at the same level as the meniscus, as shown in Figure 14B.3. Read the numbers corresponding to the lowest level of the darkened area of the meniscus. Record the initial volume reading.

Preparation of the Milk Sample

1. Obtain three 250-mL Erlenmeyer flasks. Make sure they are clean and rinsed with distilled water. They do not have to be dry.

2. Obtain roughly 60 mL of milk in a clean, dry beaker.

3. Take a 10.0-mL pipet, and check to see if it is clean by rinsing distilled water through it. Rinse the inside with some milk three times, by drawing a little milk up into the pipet with a pipet pump or bulb, tilting the pipet to rinse the sides, and then releasing the milk into a container, as shown in Figure 14B.4.

4. Using the 10.0-ml pipet and a bulb or other device, deliver 10.0 mL of milk into each Erlenmeyer flask. Label them 1, 2, and 3.

5. To each Erlenmeyer flask add approximately 40 mL of distilled water using a graduated cylinder. The amount of water used here does not affect the amount of calcium detected.

6. Add 2 mL of buffer solution to each flask.

7. Add a spatula tip full of Eriochrome Black T to each flask. It is important to add only a very small amount of the indicator since the fainter the end point,

Figure 14B.4
The correct procedure for using a pipet with a rubber suction bulb. Always use a bulb for suction—*never by mouth!*

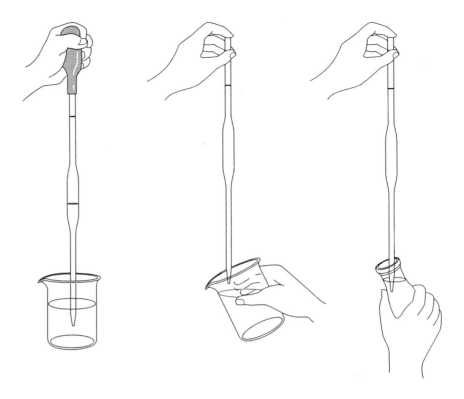

the easier it is to detect. You can always add more indicator later if it is too faint for you.

8. Mix by swirling each flask.

The Titration

1. Place flask 1 under the buret. Make sure your initial volume reading from the buret is still the same. If not, you may want to obtain a new buret since yours most likely has a leak.

2. The first titration gives a general estimate of how much EDTA is required to reach the end point. It should go relatively quickly. Add the EDTA solution 1 ml at a time. Swirl the flask as you add the EDTA solution. The color of the solution in the flask will change from lavender to blue. When the solution shows signs of blue color that persist longer during swirling, the end point is near. When the faintest blue color stays, the end point has been reached. Touch the inside of the flask to the tip of the buret. Rinse this drop down with distilled water.

3. Read and record the level of the meniscus to two places after the decimal (the hundredth place). You will have to estimate the last digit to the nearest 0.05 mL.

4. Subtract the initial from the final reading, and record this on your report form.

5. Repeat steps 1–4 for flasks 2 and 3. Do these trials more carefully. Add the EDTA quickly until approximately 3 ml before the end point, and then dropwise until the end point is reached. If your trials for flasks 2 and 3 do not agree well (are not precise), see your instructor about doing another trial.

14B
Analysis of Calcium in Milk

NAME ______________________

SECTION ______________________

DATE ______________________

Prelaboratory Assignment

1. Define the following terms:
   a. buret

   b. titration

   c. indicator

   d. end point

2. How would your final answer of milligrams of calcium be affected if you added 42 mL of water to the flask containing the milk sample instead of the specified 40 mL?

3. What would be the long-term effects of drinking a glass of EDTA solution every day? (Neglect possible toxicity.)

4. How would your calcium concentration result be affected if you did not rinse your wet buret with EDTA solution before titrating?

5. A student gets an average value of 37.8 mL of EDTA solution for the titration. How many milligrams of calcium was in the 10.0-mL milk sample? Show your calculations.

6. How many ounces of whole milk does a 20-year-old woman need per day? Use the information from Tables 14B.1 and 14B.2 and show work below.

7. How many milliliters of 0.0200 M EDTA solution would be needed to titrate 1.5 cups of eggnog? Use the information from Table 14B.2 and show work below.

14B
Analysis of
Calcium in Milk

NAME _____

SECTION _____

DATE _____

Results

Milk sample no. _____

	Flask 1 (trial run)	Flask 2	Flask 3
Final buret reading	_____	_____	_____
Initial buret reading	_____	_____	_____
Volume of EDTA	_____	_____	_____
Average volume of EDTA	_____		

Questions

1. The number of milligrams of calcium in your 10.0-mL sample using your average volume of EDTA solution was _____ mg. Show work below, including units.

2. The number of milligrams of calcium in an 8.00-oz serving of your milk sample was _____ mg. Show work below, including units.

3. How well do your results agree with the values listed in Table 14B.2? Back up your written response to this question by finding the percentage error. Show work below.

4. If you titrated a tablespoon of low-fat yogurt with 23 mg of calcium, what volume of 0.0190 M EDTA solution would you expect to react?

5. If milk contains 312 mg calcium per glass and you drink 4 glasses per day, are you necessarily getting the required amount of calcium? (Assume you require 1000 mg Ca^{2+} per day.) Why or why not?

15
Le Chatelier's Principle

Objective

- Investigate how Le Chatelier's Principle can be used to predict and explain the direction of shift in equilibrium caused by a change in species concentration and a change in temperature.

Introduction

Many chemical reactions are reversible, that is, they may proceed in both the **forward direction** (the "original" reaction where *reactants* are transformed to *products*) and the **reverse direction** (products reacting to re-form reactants). For example, in the general reaction where A and B are reactants and C and D are products:

$$A + B \underset{\text{Reverse}}{\overset{\text{Forward}}{\rightleftharpoons}} C + D$$

The forward and reverse reactions occur simultaneously. Initially, when the concentration of starting material, or reactants, ([A] and [B]) is high, the forward reaction rate is very fast. As product (C and D) is produced, the concentration of reactant material is diminished, and the forward reaction rate decreases. At the same time the reverse reaction rate increases as the concentration of product ([C] and [D]) increases. When the rate of the forward reaction exactly equals the rate of the reverse reaction, the system is said to be at **equilibrium**. The concentrations of all the species (reactants and products) remain constant when the system is at equilibrium. It is important to remember, however, that a system at equilibrium is still in a dynamic state because both the forward and reverse reactions are still occurring.

When the concentration of any species in the equilbrium system is changed, the equilibrium is disturbed and the system reacts so as to minimize the effect of the disturbance. This is known as **Le Chatelier's Principle**. For example, consider the equilibrium:

$$NH_3(aq) + H^+(aq) \rightleftharpoons NH_4^+(aq)$$

Suppose more acid (H^+) is added, thereby increasing the concentration of H^+. The system shifts in the direction to counteract this change—that is, the forward reaction rate increases temporarily to reduce the concentration of H^+. We say the system "shifts in the forward direction" to consume H^+. Similarly, if more NH_4+ is added, the system shifts in the direction to consume NH_4^+, and the reverse reaction increases. If base (OH^-) is added to the reaction vessel, it consumes the reactant acid (H^+), with the end effect of decreasing H^+ concentration. Then the system shifts to increase H^+ concentration, so the reverse direction increases.

Evidence of a shift in equilibrium is easily observed if one or more of the species are colored or if a change in phase, such as precipitation or dissolution, accompanies the shift. During this experiment you will observe the effect of changing the concentration of chemical species involved in four different equilibrium systems. You will then apply Le Chatelier's Principle to explain the observed shifts.

Experimental

Equipment: 7 small (13 mm × 100 mm) test tubes
filter paper
10-mL graduated cylinder
100-mL graduated cylinder
2 100-mL beakers
medicine droppers

Materials: distilled water
0.1 M Fe(NO$_3$)$_3$
0.1 M KSCN
0.1 M Ni(NO$_3$)$_2$
1.0 M Ca(NO$_3$)$_2$
0.1 M HC$_2$H$_3$O$_2$
0.1 M Co(NO$_3$)$_2$
6.0 M NH$_3$ [aqueous ammonia is also known as
ammonium hydroxide, NH$_4$OH(aq)]
6.0 M NaOH
6.0 M HCl
concentrated HCl
NaC$_2$H$_3$O$_2$
methyl orange indicator

Safety Precautions

Solutions of ammonia, hydrochloric acid, and sodium hydroxide can cause chemical burns in addition to ruining your clothing. Do *not* use your finger as a stopper when mixing these solutions. If you spill any of these solutions on you, wash the contaminated area thoroughly and report the incident to your laboratory instructor. You may require further treatment.

Procedure

Equilibrium of Fe^{3+} and SCN$^-$ with Fe(SCN)$^{2+}$

1. Add 20 mL of distilled water from a graduated cylinder to a 100-mL beaker. Next add 20 drops of 0.1 M Fe(NO$_3$)$_3$ and 20 drops of 0.1 M KSCN to the beaker. The iron(III) thiocyanate ion can be formed according to the equation:

$$Fe^{3+}(aq) + SCN^-(aq) \rightleftharpoons Fe(SCN)^{2+}(aq)$$

2. Using a 10-mL graduated cylinder, add 3 mL of this solution to three small test tubes marked 1, 2, and 3.
3. Add 20 drops of 0.1 M Fe(NO$_3$)$_3$ to test tube 1 and mix.
4. Add 20 drops of 0.1 M KSCN to test tube 2 and mix.

5. Add 20 drops of distilled water to test tube 3 and mix. The color of the contents of this test tube will serve as a reference.

6. Compare the colors in test tubes 1 and 2 with the color of the reference test tube. The intensity of the color in each test tube indicates the relative concentration of $Fe(SCN)^{2+}$. For best results, view the test tubes against a white paper. Record your observations.

Equilibrium of Ni^{2+} and $NH_3(aq)$ with $Ni(NH_3)_6^{2+}$

1. Add 10 drops of 0.1 M $Ni(NO_3)_2$ to a clean test tube. Record the color.

2. Add $NH_3(aq)$ dropwise until the color changes and intensifies. Record the color. The nickel–ammonia ion can be produced according to the following equation:

$$Ni^{2+}(aq) + 6NH_3(aq) \rightleftharpoons Ni(NH_3)_6^{2+}(aq)$$

3. Add 6.0 M HCl dropwise until the color changes once again. Record the color. The acid has reacted with $NH_3(aq)$ to form NH_4^+ ions.

Ionization of Acetic Acid

1. Consider the equilibrium:

$$HC_2H_3O_2(aq) \rightleftharpoons C_2H_3O_2^-(aq) + H^+(aq)$$

None of the reactants or products is colored. Therefore an indicator, methyl orange, must be added to permit detection of any shift in the equilibrium.

2. Pour 2–3 mL of 0.1 M acetic acid, $HC_2H_3O_2$, into a test tube, and add 2 or 3 drops of methyl orange. Record the color.

3. Place a few crystals of sodium acetate, $NaC_2H_3O_2$, in the solution, and shake gently to dissolve them. Record the color.

4. Pour 2–3 mL of 0.1 M acetic acid into a clean test tube, and add 2 or 3 drops of methyl orange. Add a few drops of 6.0 M NaOH. Record the color.

Solubility of $Ca(OH)_2$

1. Using a 10-mL graduated cylinder, add 5 mL of 6.0 M NaOH to a clean beaker. Rinse the graduated cylinder, then use it to add 5 mL of 1.0 M $Ca(NO_3)_2$ to the same beaker.

2. Stir the mixture with a stirring rod. A white precipitate of $Ca(OH)_2$ should be observed.

3. Using gravity filtration, filter the mixture. This filtration may require a long time. While you are waiting, you may wish to answer some of the Questions.

4. Wash the precipitate on the filter paper with 5 mL of distilled water.

5. With a metal spatula, remove as much of the wet precipitate from the filter paper as you can. Place this solid in a clean beaker along with 10 mL of distilled water.

6. Add 2 mL of 6.0 M HCl, and stir the contents of the beaker thoroughly. Record the results.

7. Add 5 mL of 6.0 M NaOH to the beaker. Record the results. You should be able to deduce the identity of the substance that has formed.

Equilibrium of Co(H₂O)₆²⁺ and Cl⁻ with CoCl₄²⁻ and H₂O

Consider the equilibrium:

$$Co(H_2O)_6^{2+}(aq) + 4\ Cl^-(aq) \rightleftharpoons CoCl_4^{2-}(aq) + 6\ H_2O(l)$$

1. Place a clean beaker containing water on a heating plate, and heat the water to a gentle boil.
2. Add 5 drops of 0.1 M Co(NO₃)₂ to a clean test tube. Record the color. This is the characteristic color of $Co(H_2O)_6^{2+}$.
3. Add 5 drops of concentrated HCl (*not* 6.0 M HCl). Shake the test tube gently, and record the color. This is the characteristic color of $CoCl_4^{2-}$.
4. Add 5 drops of distilled water. Shake gently. Record the color.
5. Place the test tube in boiling water, and wait a few minutes until the color has changed again. Record the color.
6. Cool the test tube in cold water or ice until the color changes once more. Record the color.

16A

How Much Acetic Acid Is in Vinegar?

Objective

- Determine the concentration of acetic acid in white vinegar by titration with a base, sodium hydroxide.

Introduction

The objective of this experiment is to determine the molarity (molar concentration) of acetic acid ($HC_2H_3O_2$) in vinegar. You will accomplish this through the titration of a sample of vinegar with a solution of sodium hydroxide (NaOH). These substances react according to the following equation:

$$HC_2H_3O_2(aq) + NaOH(aq) \longrightarrow NaC_2H_3O_2(aq) + H_2O(l)$$

Phenolphthalein will be used as the indicator in this titration, because it is colorless before the completion of the reaction but pink after the completion. Be prepared to observe carefully the point in the titration at which one drop of the NaOH solution causes the solution being titrated to turn from colorless to a barely discernible pink color. This point is the **end point**. You will do a trial titration to find the approximate end point before you do a pair of exact titrations.

During the experiment you will titrate a carefully measured volume of the acid with NaOH solution. The product of the volume of NaOH times its molarity is the number of moles of NaOH consumed in the reaction. This value is converted to moles of acetic acid from the stoichiometry of the reaction. The number of moles of acid divided by the volume of the vinegar sample is the molarity of the acetic acid.

Experimental

Equipment: 2 250-mL Erlenmeyer flasks and 1 stopper
small beaker
100-mL graduated cylinder
10-mL transfer pipet
50-mL buret
plastic wash bottle

Materials: white 5% vinegar
0.2–0.3 M NaOH
phenolphthalein

Safety Precautions

Never use your mouth to draw liquid into the pipet. Use a rubber suction bulb.

Procedure

Getting Started

1. Obtain a 10-mL transfer pipet and a 50-mL buret.
2. Acquire about 30 mL of white 5% vinegar and about 85 mL of the NaOH solution. The vinegar may be kept in a clean, dry beaker, but the NaOH solution must be kept in a clean, dry Erlenmeyer flask sealed with a rubber stopper. The NaOH solution must be in a closed container because NaOH reacts with carbon dioxide (CO_2) in the air. The molarity of the NaOH solution is given on the bottle label and will be between 0.2 and 0.3 M.
3. The Introduction to this manual gives instructions for using a buret. After you have read those procedures carefully, clean your buret and fill it with the NaOH solution.

Trial Titration

1. Pipet 10.0 mL of the vinegar into a clean Erlenmeyer flask. Add about 20 mL of distilled water from a clean graduated cylinder. Add 2 drops of phenolphthalein solution.
2. Record the molarity of the NaOH solution and the initial buret reading. (See Figure 14B.3 for the correct method.) Remember to record the volumes read from the buret to the nearest 0.05 mL.
3. Place the flask under the buret with the capillary tip inside the mouth of the flask. Place a piece of white paper under the flask.
4. Add the NaOH solution to the flask in increments of about 1 mL while swirling. Note the color of the solution after each addition.
5. This trial titration is complete when an addition of about 1 mL causes the color to change from colorless to any shade of pink, and the pink color is retained even after swirling the solution.
6. Record the buret reading. Subtract the initial reading from this final reading to obtain the volume required for the approximate end point.

Exact Titrations

1. Repeat steps 1, 2, and 3 of the procedure used for the trial titration.
2. Subtract 1 mL from the total volume required to obtain the end point found in the trial titration. Rapidly add the resulting volume to the flask from the buret.
3. Rinse the walls of the Erlenmeyer flask with distilled water from a plastic wash bottle.
4. Continue the titration on a drop-by-drop basis. Swirl the flask rapidly after each drop. The end point is the first permanent, barely visible pink color. Finding the true end point requires patience and skill. Absolutely no skill is required to miss the end point and achieve a very deep pink color. If you are unsure about the end point, record the buret reading before you add the next drop.
5. Repeat the procedure with a second sample of vinegar.

6. If the volumes at the end points of the two exact titrations differ by more than 0.15 mL (about 3 drops), repeat the titrations with additional samples of vinegar until two consecutive results have this precision.
7. Calculate and record the molarity of the vinegar from each titration. Also, calculate the average value of molarity.

16B

pH: What Really Is the Color of Red Cabbage?

Objectives

- Determine the color of an indicator at various pH values.
- Determine the pH and acidity of some common household solutions.

Introduction

Acids and bases are a part of our everyday existence. For example, acid rain has been shown to damage not only buildings and monuments (as shown in Figure 16B.1) but also trees and aquatic life. Your shampoo is probably pH balanced, and almost all cleaning supplies such as window cleaners, soaps, and cleansers are basic, or alkaline. A car's battery contains sulfuric acid. During this experiment you will get an opportunity to discover how acidic or basic some of your cosmetics and commercial cleaning and food products are. Now let us define some of these words related to acidity.

An **acid** is a substance that can donate a hydrogen cation, H^+. For example, OH^- will accept (react with) a hydrogen cation to form water. (The hydrogen cation, H^+, is simply a hydrogen atom without the electron; hence it is sometimes referred to as a proton.) An acid can be easily recognized by its formula, which usually has H^+ at the beginning. Some examples are: HCl, hydrochloric acid; H_2SO_4, sulfuric acid; H_3PO_4, phosphoric acid; and HNO_3, nitric acid.

A **base** is a substance that can accept a hydrogen cation, H^+, or form OH^- in water. A base usually can be recognized by its formula, which has hydroxide ion at the end. Some examples are: NaOH, sodium hydroxide; $Ca(OH)_2$, calcium hydroxide; and NH_4OH, ammonium hydroxide.

Acids and bases come in a variety of strengths, relating to their ability to release (for acids) or accept (for bases) the H^+. A strong acid is one that tends to release H^+ completely or almost completely. It is said to completely *dissociate*. Nitric acid, sulfuric acid, and hydrochloric acid are all examples of a strong acid.

Figure 16B.1
Effect of acid rain. The deterioration of this ancient statue is blamed on acid rain and air pollution.

A weak acid is one that tends to dissociate only to a small extent, that is, it tends to hold onto its H^+ most of the time. Acetic acid, $HC_2H_3O_2$, the substance that gives vinegar its tart flavor and characteristic odor, is considered a weak acid. Similar descriptions apply to strong and weak bases.

Besides strengths, acids and bases come in a variety of concentrations, usually expressed in terms of molarity (M), or moles/liter (mol/L). *Dilute* and *concentrated* are another common way to describe acids and bases.

pH is a measure of acidity; it is affected by both acid strength and acid concentration. Mathematically,

$$pH = -\log[H_3O^+]$$

where $[H_3O^+]$ represents the concentration of H_3O^+ ions in units of molarity. Consider these H_3O^+ concentrations and pH values:

H_3O^+ concentration (M)	pH
$1 = 1 \times 10^0$	0
$0.1 = 1 \times 10^{-1}$	1
$0.01 = 1 \times 10^{-2}$	2
$0.00001 = 1 \times 10^{-5}$	5
1×10^{-8}	8
1×10^{-10}	10
1×10^{-14}	14

Do you notice the relationship between the $[H_3O^+]$ exponent and the pH value?

The lower the pH value, the greater the acidity of the solution. The higher the pH value, the less acidic and the more basic is the solution. (Note the apparent contradiction: a *low* pH means a *strong* acid.) In an aqueous solution both H_3O^+ and OH^- are present at all times. This is due to the equilibrium between H_2O and these ions, referred to as the *self-ionization of water*. The equation for this equilibrium is:

$$2H_2O(l) \rightleftharpoons H_3O^+(aq) + OH^-(aq)$$

In a neutral solution the pH is 7, and the concentration of both H_3O^+ and OH^- is $1 \times 10^{-7}M$. If a large amount of H_3O^+ is present, the concentration of OH^- decreases. If the concentration of H_3O^+ decreases, the concentration of OH^- increases. This relationship can be obtained from the equilibrium expression above and is expressed mathematically as:

$$[H_3O^+] \times [OH^-] = 1.0 \times 10^{-14}$$

It is sometimes useful to find the OH^- concentration, and the formula above can be used. For example, if a solution has a $[H_3O^+] = 2.5 \times 10^{-5}$ M, the OH^- concentration could be calculated by using this formula:

$$[OH^-] = \frac{1.0 \times 10^{-14}}{[H_3O^+]} = \frac{1.0 \times 10^{-14}}{2.5 \times 10^{-5}} = 4.0 \times 10^{-10} \, M$$

In a similar manner, we can calculate pOH values. The mathematical definition of pOH is analogous to that of pH:

$$pOH = -\log[OH^-]$$

We can also find pOH values using the following relationship:

$$pH + pOH = 14$$

Several different experimental methods may be used to determine pH. One instrument, known as a *pH meter,* measures pH electronically. Another common method, which will be used in this experiment, employs an **indicator**. An indicator is a compound whose color is pH dependent. For example, phenolphthalein is colorless at pH values lower than 8 and pink at pH values greater than 8. There are many indicators that change color at a particular pH value. Other indicators change color many times over a broad pH range. These are known as **universal indicators**. For example, a universal indicator may be blue at pH 7, green at pH 8, yellow at pH 9, orange at pH 10, and so forth.

The purple compound in red cabbage is a universal indicator. You will extract (remove) this indicator from the cabbage, determine its color at various pH standards, and then use it to determine the pH of common household items.

Experimental

Equipment: 250-mL beaker
100-mL beaker
10-mL graduated cylinder
balance
heating plate
thermometer
funnel
ring stand, iron ring
filter paper
test tube rack
22 small (13 mm × 100 mm) test tubes
marking pencil

Materials: red cabbage leaves
distilled water

1.0 M KCl
12 solutions of pH 1 through 12
10 household substances

Procedure

Extraction of the Indicator

1. Obtain about 12 g of red cabbage leaves. Tear them into small, stamp-sized pieces, and place them in a 250-mL beaker.
2. Add approximately 70 mL of distilled water to the cabbage. Using a heating plate, heat the beaker with the cabbage to just under boiling, approximately 90° to 95°C. Check occasionally with a thermometer. DO NOT BOIL!
3. Allow the beaker with the cabbage to cool. Set up the ring stand with an iron ring, funnel, and filter paper, as shown in Figure I.6 (page 11).
4. Decant the cabbage solution into the filter paper as shown in Figure I.7 (page 12). Save the solution, and dispose of the leaves in the appropriate waste receptacle.

Determination of Indicator's Color at Various pH Values

1. Obtain a test tube rack and 12 small test tubes. Label the test tubes 1 through 12, to represent the pH values you will use.
2. Using a 10-mL graduated cylinder, add 2 mL of the indicator to each test tube.
3. Add 3 drops of 1.0 M KCl solution to each test tube. Mix the solutions gently.
4. Locate the standard pH solutions, and add 10 drops of pH 1 to test tube 1, 10 drops of pH 2 to test tube 2, and so on.
5. Record the colors of the solutions in the test tubes. Save these solutions for the duration of this experiment.

Determination of pH Values of Household Substances

1. Add 2 mL of indicator (the decanted cabbage solution) to 10 test tubes. Also add 3 drops of 1.0 M KCl solution to each test tube.
2. To each test tube add a different household substance, including tap water. Follow any special procedures that might apply (see a–d). Record the color after mixing. Compare the color to the standard solutions, and assign the closest pH value.
   a. If your substance is a liquid and *not* listed below, add 10 drops of it to a test tube.
   b. If your substance is a *viscous substance,* like shampoo, dip a stirring rod ¼ in. into the substance and use the rod to swirl the adhering substance into the test tube, which already contains the indicator and KCl solution.
   c. If your substance is a *solid,* grind up 1 tablet or 0.2 g of it and dissolve in 100 mL of distilled water. Add 10 drops of the dissolved substance to a test tube with indicator and KCl solution.

d. If your substance is one of the following, add the specified number of drops to the test tube with indicator and KCl solution.

Substance	Number of drops
vinegar	2
lemon juice	1
white wine	4
orange juice	4
Mr. Clean or other strong cleaning product	1

16B
pH: What Really Is the Color of Red Cabbage?

NAME ____________________

SECTION ____________________

DATE ____________________

Prelaboratory Assignment

A reminder: You may want to bring several clear or colorless items to lab this week to be tested for pH. Some suggestions: fruit juice, soda, shampoo, cooking or cleaning supplies, analgesics (aspirin, acetaminophen, etc.).

1. Define the following terms:

   acid

   base

   neutral

   pH

   indicator

2. Classify each of the following as an acid or base or neutral (neither an acid nor a base). See the Introduction to this experiment for some examples.

   NH_4OH $CaCl_2$ H_3PO_4 $NaCl$ $Sr(OH)_2$ HF

3. In the following table, fill in the [H⁺], pH, and a description of the solution as strongly acidic, weakly acidic, neutral, weakly basic, or strongly basic.

$[H_3O^+]$	$[H_3O^+]$ in Exponential Notation	pH	Description
0.00001 M	_____	_____	_____
0.1 M	_____	_____	_____
_____	1×10^{-7} M	_____	_____
_____	_____	10	_____
_____	1×10^{-3} M	_____	_____

4. If the $[H_3O^+]$ of a solution is 0.00000000001 M, what are its pH, $[OH^-]$, and pOH? Show all work.

16B
pH: What Really Is the Color of Red Cabbage?

Results

1. *Standard solutions*

pH	Description of Color with Indicator
1	_____
2	_____
3	_____
4	_____
5	_____
6	_____
7	_____
8	_____
9	_____
10	_____
11	_____
12	_____

2. *Household items*

	Substance	Description of Color	pH
1	_____	_____	____
2	_____	_____	____
3	_____	_____	____
4	_____	_____	____
5	_____	_____	____
6	_____	_____	____
7	_____	_____	____
8	_____	_____	____
9	_____	_____	____
10	_____	_____	____

Questions

1. From your pH values of household substances, which substance was the closest to a neutral pH? _____

   Which was most acidic? _____

   Which was most basic? _____

2. From your pH values, list your household items in order of *increasing* acidity. (List the most acidic one at the bottom and the least acidic at the top.)

3. Complete the following table:

$[H_3O^+]$	pH	pOH	$[OH^-]$
_____	1	_____	$1 \times 10^{-13}\ M$
$1 \times 10^{-4}\ M$	_____	_____	_____
_____	9	_____	_____
_____	_____	5	_____
_____	_____	14	_____
_____	6	_____	_____
_____	_____	0	_____

4. Using your data and the pH that corresponds to the color closest to yellow, find the $[OH^-]$.

5. For your brightest pink solution and matching pH, calculate the pOH.

17

Oxidation–Reduction Reactions

Objectives

- Determine experimentally the relative strengths of several metals as reducing agents.

- Determine experimentally the relative strengths of several halogens as oxidizing agents.

Introduction

Oxidation–reduction reactions involve the transfer of electrons. Consider the following net ionic equation for the oxidation–reduction (or redox) reaction between magnesium metal and copper cations to form magnesium cations and copper metal:

$$Mg(s) + Cu^{2+}(aq) \longrightarrow Mg^{2+}(aq) + Cu(s)$$

This reaction may be separated into two half-reactions, an **oxidation half-reaction**:

$$Mg \longrightarrow Mg^{2+} + 2e^-$$

and a **reduction half-reaction**:

$$Cu^{2+} + 2e^- \longrightarrow Cu$$

Note that in the oxidation half-reaction there is a loss of electrons, that is, electrons are one of the products of the reaction. In the reduction half-reaction there is a gain of electrons; electrons are one of the reactants. Adding the two half-reactions together yields the oxidation–reduction reaction.

Observe that the number of electrons lost by magnesium in the oxidation half-reaction is exactly equal to the number of electrons gained by copper in the reduction half-reaction. This balance is essential in any oxidation–reduction reaction; there can never be "left-over" electrons as either reactants or products. Therefore, to balance an oxidation–reduction reaction, the two half-reactions must be adjusted so that an equal number of electrons is gained and lost during the reaction.

Let us consider the following hypothetical reaction where metal A is oxidized to become cation A^{3+}:

$$A \longrightarrow A^{3+} + 3e^- \qquad \text{Oxidation half-reaction}$$

and B, a halogen (and therefore existing as a diatomic molecule, B_2), is reduced:

$$B_2 + 2e^- \longrightarrow 2B^- \qquad \text{Reduction half-reaction}$$

To write the balanced oxidation–reduction reaction between A and B_2, we multiply the oxidation half-reaction by 2 and the reduction half-reaction by 3 to equalize the electrons gained and lost.

$$
\begin{array}{rcl}
2A & \longrightarrow & 2A^{3+} + 6e^- \\
3B_2 + 6e^- & \longrightarrow & 6B^- \\
\hline
2A + 3B_2 & \longrightarrow & 2A^{3+} + 6B^-
\end{array}
$$

In this example, metal A essentially gives its electrons to B_2, that is, metal A causes halogen B_2 to be reduced. Consequently, we refer to metal A as a **reducing**

agent. It causes reduction to occur but is itself oxidized. On the other hand, halogen B_2 obtains electrons from metal A. We refer to B_2 as an **oxidizing agent**. An oxidizing agent causes oxidation of the other reactant but is itself reduced. In an oxidation–reduction reaction, the oxidizing agent is always reduced, and the reducing agent is always oxidized.

In general, metals and negatively charged simple ions tend to lose electrons and therefore typically are reducing agents (although there are exceptions such as OCl^-). Elemental halogens (Cl_2, Br_2, I_2) and positively charged simple ions tend to gain electrons and typically are oxidizing agents (although there are exceptions such as NH_4^+).

Oxidizing and reducing agents vary in strength, depending on their tendency to capture or release electrons. For example, a strong oxidizing agent has a great affinity for electrons and causes many other chemical species to be oxidized. The strongest oxidizing agent is the one that causes the greatest number of oxidation half-reactions to occur.

In this experiment you will examine the relative strengths of a number of oxidizing and reducing agents. First you will mix combinations of metals and metal ions to obtain your results. The half-reactions you will examine (in alphabetical order, and written by convention as reductions) are:

$$Ag^+ + e^- \longrightarrow Ag$$
$$Cu^{2+} + 2e^- \longrightarrow Cu$$
$$Pb^{2+} + 2e^- \longrightarrow Pb$$
$$Zn^{2+} + 2e^- \longrightarrow Zn$$

You will analyze your data and arrange these half-reactions in order of decreasing strength of the reducing agents.

In the second part of the experiment you will examine the reactions of several halogens. The elemental halogens—Br_2, Cl_2, and I_2—are all relatively strong oxidizing agents. Upon reduction, each forms halide ions:

$$Br_2 + 2e^- \longrightarrow 2Br^-$$
$$Cl_2 + 2e^- \longrightarrow 2Cl^-$$
$$I_2 + 2e^- \longrightarrow 2I^-$$

You will arrange the halogens in order of increasing strength of the oxidizing agents. To accomplish this, you will mix combinations of halogens with the other halides and determine whether a reaction has occurred.

You will determine whether a reaction has occurred by extracting the mixture of halogens and halides with cyclohexane, an organic solvent that is immiscible with water. In each case the elemental halogen (I_2, Br_2, or Cl_2) in the aqueous solution will be extracted into the cyclohexane, where its characteristic color may be observed. Prior to mixing the various combinations, you will extract aqueous solutions of Cl_2, Br_2, and I_2 for standards of comparison.

Consider this example. You mix chlorine (Cl_2) with iodide ion (I^-) to see whether a reaction occurs. Initially you note that free chlorine, Cl_2, is colorless in the cyclohexane layer. When the chlorine and iodide are mixed:

$$Cl_2 + 2I^- \longrightarrow ?$$

the cyclohexane layer changes color to light pink so you know that iodine, I_2, is one of the products, and therefore that a reaction has occurred. This is a hypothetical example that is intended to show how the method works. However, reality may be entirely different.

Experimental

Equipment: 6 small (13 mm × 100 mm) test tubes
emery (or sand) paper
medicine dropper

Materials: 0.1 M $Zn(NO_3)_2$
0.1 M $Pb(NO_3)_2$
0.1 M $Cu(NO_3)_2$
0.1 M $AgNO_3$
clean strips of copper, zinc, and lead
$Br_2(aq)$ (bromine water)
$Cl_2(aq)$ (chlorine water)
$I_2(aq)$ (iodine water)
0.1 M NaBr
0.1 M NaCl
0.1 M NaI
cyclohexane
Kimwipes

Safety Precautions

Bromine and chlorine waters release vapors that are extremely harmful when inhaled. Cyclohexane is also harmful and should not be inhaled. Handle these liquids only in a fume hood and with utmost care.

Do not pour solutions containing cyclohexane down the drain. Dispose of them in a proper waste bottle.

Procedure

Metals as Reducing Agents

1. Obtain one strip each of copper, zinc, and lead. Clean one side of each strip with sand or emery paper. Lay the strips on a paper towel with the cleaned surface up.

2. Place on the copper strip one drop of $Zn(NO_3)_2$, one drop of $Pb(NO_3)_2$, and one drop of $AgNO_3$. Make sure the drops do not touch one another. Allow 5 min before you decide whether a reaction has occurred. An easy way to test for a reaction is to rub the spot where the drop was on the metal strip with a Kimwipe. There will be a black residue (due to the metal plated out on the strip) on the Kimwipe if a reaction has occurred. Record your observations.

3. Repeat step 2 for zinc and lead. Place one drop of $Cu(NO_3)_2$, one drop of $Pb(NO_3)_2$, and one drop of $AgNO_3$ on the zinc strip. Place one drop of $Cu(NO_3)_2$, one drop of $Zn(NO_3)_2$, and one drop of $AgNO_3$ on the lead strip. Record your observations.

4. Return the strips to their proper location.

5. Write the net ionic equations for all reactions that actually occurred.

6. Rank the metal cations in order of increasing stength as oxidizing agents.

7. Rank the metals in order of decreasing stength as reducing agents. Include silver in your ranking by assuming that silver metal does not react with Zn^{2+}, Cu^{2+}, or Pb^{2+}.

Halogens as Oxidizing Agents

1. Arrange six small test tubes in a rack and label them 1 through 6.

2. Add half a medicine dropperful of cyclohexane into each test tube.

3. Add a few drops of chlorine water to test tubes 1 and 2. Add a few drops of bromine water to test tubes 3 and 4. Add a few drops of iodine water to test tubes 5 and 6.

4. Shake the solutions and record the initial color of each.

5. Add 8 to 10 drops of halide solution to the test tubes as shown in the following table. Shake the test tubes to mix the two solution layers. Then allow the two phases to separate. Water and cyclohexane are immiscible, so cyclohexane, the less dense liquid, rises to the top, carrying any free halogen (Cl_2, I_2, or Br_2) with it. By the color of this top layer, determine which elemental halogen is present after mixing and thus whether a redox reaction has taken place. Has the initial color (from step 4) changed? If so, what does it mean? Record your observations.

6. Examine the results of your tests, and write the net ionic equations for each reaction that occurred.

7. Arrange the halogens in order of increasing strength as oxidizing agents.

8. Arrange the halide ions in order of decreasing strength as reducing agents.

Test Tube	Halogen	Halide Solution
1	Cl_2	NaBr
2	Cl_2	NaI
3	Br_2	NaCl
4	Br_2	NaI
5	I_2	NaCl
6	I_2	NaBr

18
Natural Radioactivity

Objectives

- Measure the rate of disintegration of potassium-40 in an ordinary sample of KCl.

- Calculate the half-life of this isotope.

Introduction

Radioactivity is the process whereby an atomic nucleus spontaneously decomposes, creating a different nucleus and emitting one or more particles. The nuclei of radioactive isotopes become more stable during this transformation by changing the number of neutrons and/or protons to a more favorable value. One example of a radioactive substance is cobalt-60 (cobalt-60 is the isotope of cobalt with mass number 60). The particles emitted by cobalt-60 are used to irradiate fresh fruits and vegetables to inhibit growth of bacteria and fungi.

Radioactive isotopes, such as cobalt-60, carbon-14, and uranium-238, may undergo five different types of decay: (1) beta emission, (2) alpha particle emission, (3) gamma ray emission, and (4) positron emission, or (5) electron capture.

Carbon-14, used for dating ancient articles of wood, cloth, or bones, decays by **beta emisson**:

$$^{14}_{6}C \longrightarrow ^{\ \ 0}_{-1}e + ^{14}_{7}N$$

As shown in this example, the net effect of a beta particle ($^{\ \ 0}_{-1}e$) emission is to change a neutron to a proton. A beta particle is identical to an electron.

When an element emits an **alpha particle** ($^{4}_{2}He^{2+}$), its atomic number decreases by 2 and its mass number decreases by 4. An alpha particle is a helium ion with a +2 charge and therefore no electrons. Uranium-238 decays by both alpha particle and gamma ray emission:

$$^{238}_{92}U \longrightarrow ^{4}_{2}He^{2+} + ^{234}_{90}Th + 2^{0}_{0}\gamma$$

Gamma rays ($^{0}_{0}\gamma$) are short-wavelength light rays and therefore have zero charge and zero mass number. Uranium-238 was the first naturally occurring radioactive isotope discovered.

A **positron** ($^{0}_{+1}e$) has a +1 charge and the same mass as an electron. The net effect of a positron emission is to change a proton to a neutron. Sodium-22 decomposes by positron emission:

$$^{22}_{11}Na \longrightarrow ^{0}_{+1}e + ^{22}_{10}Ne$$

Radioactive decay by electron capture does not occur very often. **Electron capture** involves the capture of an inner-orbital electron by the nucleus. Gamma rays are always produced along with electron capture:

$$^{73}_{33}As + ^{\ \ 0}_{-1}e \longrightarrow ^{73}_{32}Ge + ^{0}_{0}\gamma$$

Note that in any balanced nuclear reaction the sum of the mass numbers and the sum of the atomic numbers of the reactants always equal the sum of the mass numbers and the sum of the atomic numbers of the products.

Radioactivity levels are commonly measured with a Geiger–Müller counter, or **Geiger counter**.

The Geiger counter contains argon gas. High-energy particles from radioactive decay collide with the argon gas and cause some of the argon atoms to become ionized. The ions and electrons formed during this process allow the flow of a pulse of current, which the Geiger counter detects and "counts" as a decay event.

Scientists express the rate of decay of a radioactive isotope in terms of its **half-life**. The half-life is the time required for half of any given quantity of a radioactive isotope to decay. The half-life of some radioactive elements is a matter of seconds; for example, oxygen-19 has a half-life of 29.4 s. Other elements have extremely long half-lives; for example, uranium-238 has a half-life of 4.5×10^9 years. In other words, it takes only 29.4 s for half of a given sample of oxygen-19 to decay, whereas it takes 4.5 *billion* years for the same size sample of uranium-238 to decay to half its original amount.

There are both natural and man-made sources of radioactivity. The natural sources of radioactivity have been around since long before humans existed. All of us are exposed to radioactivity from natural sources. Potassium-40, a natural source of radioactivity, is the only isotope of the three that make up potassium that is radioactive. It undergoes radioactive decay by beta emisson, electron capture, and positron emission. The half-life of potassium-40 is 1.28×10^9 years. Since the average human body has 140 g of potassium within it, all of us are slightly radioactive!

From this experiment you will be able to calculate the half-life of potassium-40 in your sample.

Experimental

Equipment: Geiger counter
balance (electronic, triple beam or platform)
100-mL beaker
clamp and iron stand

Materials: KCl (potassium chloride)

Procedure

Getting Started

1. Your instructor may want you to work with a partner.
2. Turn on your Geiger counter, and let it warm up for a least 5 min. Remove the protective shield from the counter if one is present.

Performing the Experiment

1. Measure and record the background radiation by reading the meter 10 times at intervals of 10 s. Do not turn off the Geiger counter between readings.
2. Use a balance to measure the mass of a 100-mL beaker. Record your results.
3. Add approximately 10 g of KCl to the beaker, and measure the combined mass. Record your result, and calculate the mass of your KCl sample.

Figure 18.1
Arrangement of the
Geiger counter
within the beaker
containing the sample
of KCl.

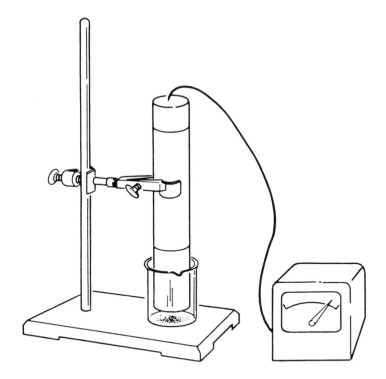

4. As shown in Figure 18.1, clamp the counter as close to the sample in the beaker as possible without it actually touching the KCl.
5. Measure and record the count rate 10 times at intervals of 10 s. Do not turn off the Geiger counter between readings.

18 Natural Radioactivity

NAME ______________________

SECTION ______________________

DATE ______________________

Prelaboratory Assignment

1. Define the following terms:
   a. radioactivity

   b. isotope

   c. half-life

   d. beta emission

   e. positron emission

   f. alpha particle emission

2. Write the nuclear equations that describe the three modes of decay for potassium-40.

3. What is cobalt-60 used for?

4. What is the function of argon in the Geiger counter?

NAME _____

SECTION _____

DATE _____

Results

Background activity (counts/s)

_____   _____   _____   _____   _____

_____   _____   _____   _____   _____

Mass of KCl and beaker (g)   _____

Mass of beaker (g)   _____

Mass of KCl (g)   _____

Activity of background and KCl (counts/s)

_____   _____   _____   _____   _____

_____   _____   _____   _____   _____

Questions

Show all work!

1. What is the average value of the background activity measured?

2. What is the average value of the total (background and KCl) activity measured?

3. What is the average value of the KCl activity alone?

4. How many potassium nuclei are in your sample of KCl?

5. There is only 0.012% of the potassium-40 isotope in any sample of potassium. Using this information and the answer to Question 4, how many potassium-40 nuclei are in your sample?

6. Calculate the half-life of potassium-40 using the following equation:

$$t_{1/2} = (0.693/\text{rate})N_t$$

where $t_{1/2}$ is the half-life of potassium-40, rate is the average value of the KCl activity (calculated in Question 3) × 5 (multiply by 5 because your counter will only detect about $\frac{1}{5}$ of the emitted radiation), and N_t is the number of potassium-40 nuclei in your sample, as calculated in Question 5. Be sure to express $t_{1/2}$ in units of *years*.

7. Calculate the percentage error when comparing your value with the known value of the half-life of potassium-40.

19
Preparation and Properties of Soap

Objectives

- Starting with animal lard, prepare a soap.
- Examine the chemical properties of the soap you have prepared.

Introduction

A **soap** is the sodium or potassium salt of a long-chain fatty acid (a type of organic compound). The fatty acid typically contains 12 to 18 carbon atoms. Solid soaps usually consist of sodium salts of fatty acids, whereas liquid soaps consist of potassium salts of fatty acids. Soaps are made by the reaction of a fat or oil such as cooking oil or cottonseed oil with concentrated sodium hydroxide (NaOH). Soap was probably first made by boiling lard with lye. Lard is an animal fat, and lye is a mixture of aqueous NaOH and KOH, obtained by leaching wood ashes with water.

The **saponification**, or soap-making, reaction is:

$$CH_2-O-\overset{\overset{\displaystyle O}{\|}}{C}-R_n$$
$$CH-O-\overset{\overset{\displaystyle O}{\|}}{C}-R_n \quad + \quad 3NaOH \longrightarrow$$
$$CH_2-O-\overset{\overset{\displaystyle O}{\|}}{C}-R_n$$

$$CH_2OH \qquad R_n-\overset{\overset{\displaystyle O}{\|}}{C}-O^-Na^+$$
$$CHOH \qquad +R_n-\overset{\overset{\displaystyle O}{\|}}{C}-O^-Na^+$$
$$CH_2OH \qquad R_n-\overset{\overset{\displaystyle O}{\|}}{C}-O^-Na^+$$

an oil (a glyceride or ester of fatty acids and glycerol)   glycerol   soap: sodium salts of the fatty acids

(*Note:* R_n = an alkyl (hydrocarbon) chain, n-carbons long (usually n = 12–18).

Soap molecules have a polar (charged) end that is **hydrophilic** (water-loving) and a long, nonpolar, organic "chain" that is **hydrophobic** (water-fearing). Soap is able to dissolve (actually to suspend or emulsify) oils or other organic materials in water in the following way: the polar ends of the soap molecules dissolve in water, and the nonpolar ends encase, or surround, the small globules of oil or "dirt," effectively making the oil "soluble" in water (see Figure 19.1). In this way soap helps remove dirt from things we want to clean, like clothes or skin. If the water in which the soap is dissolved contains appreciable amounts of such ions as calcium, magnesium, or iron (that is, the water is "hard"), these ions replace the sodium in the soap molecule, and the resulting compound becomes insoluble, forming "bathtub ring" or soap scum.

When we shake a mixture of oil and water vigorously, the oil forms very small droplets that disperse uniformly throughout the water. When we stop shaking the oil–water mixture, the oil and water form separate layers. If we add a few drops of oil to water containing some soap, and shake the mixture vigorously, the oil droplets disperse throughout the solution, as before. However, when we allow this mixture to stand, the oil droplets do not readily coalesce to form layers as before. Instead, we have formed an **emulsion**. An emulsion contains fine droplets of one liquid (oil) dispersed in another, immiscible liquid (water). The

Figure 19.1
An oil droplet has been emulsified by a soap solution. The hydrophobic end of the soap molecule is retained within the oil, and the hydrophilic end remains in the aqueous solution surrounding the oil droplet.

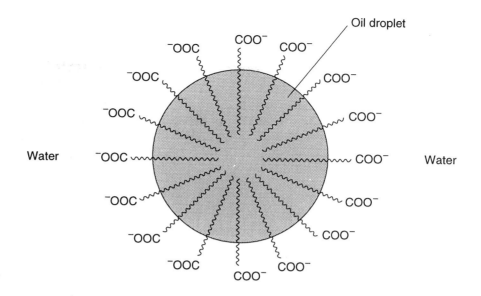

hydrophobic end of the soap molecule interacts with the nonpolar oil, and the hydrophilic end ($-COO^-$) of the soap molecule dissolves in water, as shown in Figure 19.1. The oil droplet has been emulsified by the soap solution. The better the emulsion characteristics of a soap, the better it can clean.

Experimental

Equipment: 400-mL beaker
hot plate
glass stirring rod
Büchner funnel setup
Kimwipes
paper towels
medicine dropper
5 small (13 mm × 100 mm) test tubes
rubber gloves

Materials: animal lard
ethanol (ethyl alcohol)
20% NaOH solution
saturated NaCl solution
ice (for cooling ethanol)
phenolphthalein
distilled water
5% $CaCl_2$ aqueous solution
5% $MgCl_2$ aqueous solution
5% $FeCl_3$ aqueous solution
mineral oil

Safety Precautions

Sodium hydroxide causes severe burns when in contact with the skin. If a slippery feeling is noted on the hands or other parts of the body, rinse with large amounts of cold water immediately.

Procedure

Preparation of Soap

1. Weigh a 400-mL beaker on a balance, and then weigh out 18 g of lard. Add 20 mL of ethanol and 20 mL of 20% NaOH solution. Stir the mixture.
2. Place the beaker on the hot plate and heat *gently* for approximately 30–40 minutes while stirring *constantly.* Continue heating until the odor of alcohol disappears and a *pasty mass remains* (similar to melted marshmallows). This pasty mass is the soap plus the glycerol produced in the reaction.
3. Allow the soap mixture to cool; then add 150 mL of saturated NaCl solution and stir thoroughly with a glass rod. This process is referred to as "salting out" (or precipitation) and is used to remove the soap from water, glycerol, and unreacted NaOH.
4. After the mixture has been stirred and mixed completely, filter off the soap on a Büchner funnel, using suction (see Figure I.9, on page 13 for proper setup). Use a Kimwipe instead of filter paper with the Büchner funnel. Rinse with two 10-mL portions of ice-cold ethanol while the suction is still on. Allow your soap to dry by spreading it out on a paper towel.

Properties and Reactions of Soap

1. *Washing properties.* (If you have dry or sensitive skin, you may want to wear rubber gloves for this test.) Take a small amount of your soap and wash your hands with it. In soft water, the soap should lather easily. If any oil is left over, the soap will feel greasy. Too much NaOH will also result in a slick feeling and in a roughening of your hands. Describe the washing properties of your soap on the report sheet. Rinse your hands several times after the test.
2. *Alkalinity (basicity).* A soap that contains free alkali (base) can damage your skin. To test whether your soap contains free alkali, dissolve a small piece in 5 mL of ethanol and add 2 drops of phenolphthalein. If the indicator turns red, free alkali is present. Record your observation.
3. *Hard-water reactions.* Dissolve about 1 g of your soap in 50 mL of warm distilled water. Pour about 10 mL of soap solution into each of three test tubes. To test tube 1, add 8 to 10 drops of 5% $CaCl_2$; to test tube 2, 8 to 10 drops of 5% $MgCl_2$; and to test tube 3, 8 to 10 drops of 5% $FeCl_3$. Record your observations. (Does this remind you of the "scum" that forms when you wash in hard water?)

4. *Emulsifying properties.* Put 5 to 10 drops of mineral oil in a test tube containing 8 to 10 mL of water, and shake it. A temporary emulsion or suspension of tiny oil droplets in water forms. Let this solution stand for a few minutes. Prepare another test tube with the same ingredients, but add about 0.5 g of your soap to it before shaking it. Compare the stabilities of the emulsions in the two test tubes. Which emulsion clears first? Explain. Record your answers.

NAME _____

SECTION _____

DATE _____

Prelaboratory Assignment

1. What is a soap?

2. What does the term *saponification* mean?

3. What does the term *emulsion* mean?

4. What do the terms *hydrophilic* and *hydrophobic* mean?

5. The chemical formula of a soap is:

$$K^+ \quad {}^-O\!-\!\!\underset{\displaystyle \overset{\|}{C}}{C}\!-\!C_{17}H_{35}$$

   Which end of the soap molecule is nonpolar, and which end is polar?

6. Why is the use of a soap in hard water impractical?

19
Preparation and
Properties of Soap

NAME _____

SECTION _____

DATE _____

Results

1. *Washing properties*

   Does the soap lather a lot, a little, or not at all? Does the soap feel oily?

2. *Alkalinity (basicity)*

   Soap solution + indicator

   turns pink _____   remains colorless _____

3. *Hard-water reactions*

Cation Added	Observations
Ca^{2+}	_____
Mg^{2+}	_____
Fe^{3+}	_____

4. *Emulsifying properties*

   Which mixture, oil–water or oil–water–soap, forms a more stable emulsion?

   Which emulsion clears first?

   Explanation

Questions

1. Explain how soap acts as a cleansing agent.

2. Write a chemical equation for the preparation of a soap from a fat.

3. What are two things that can cause a soap not to clean properly or to damage the skin?

20
Amylase: One of Your Enzymes

Objective

- Examine the influence of active and inactive forms of amylase on the hydrolysis of starch. Your own saliva will be the source of the enzyme.

Introduction

Starch is one of the most important sources of carbohydrates in our diet. The digestion of starch begins in the mouth. The reaction that occurs during digestion is called **hydrolysis** (reaction with water). This reaction, which really consists of a series of steps, is shown in Figure 20.1. Sugars are the ultimate products of the reaction, which could be thought of as simply breaking starch down into smaller pieces.

The hydrolysis of starch is very slow in the absence of **amylase**, an **enzyme** found in our saliva. Enzymes are proteins that are efficient and specific **catalysts** of biochemical reactions. Recall that a catalyst is a substance that speeds up a reaction without being consumed. Amylase functions most efficiently at the pH and temperature of the mouth. Extreme deviations from these conditions can cause structural changes in the enzyme that inactivate (denature) it.

The activity of salivary amylase can vary widely among different individuals and for the same individual at different times. A small percentage of individuals (perhaps 3% to 4%) may find that they are unable to do this experiment at all because their salivary amylase is inactive. Working with a partner who has active amylase is an acceptable alternative.

Throughout this experiment iodine will be used as an indicator for the presence of starch. Iodine and starch interact to yield a blue color whose intensity is proportional to the quantity of starch present. Iodine does not interact with the sugars that form as starch is hydrolyzed. During hydrolysis, the quantity of starch decreases as the reaction progresses, and the blue color fades. If the intensity of the blue color does not diminish, hydrolysis has not occurred (at least to any appreciable extent).

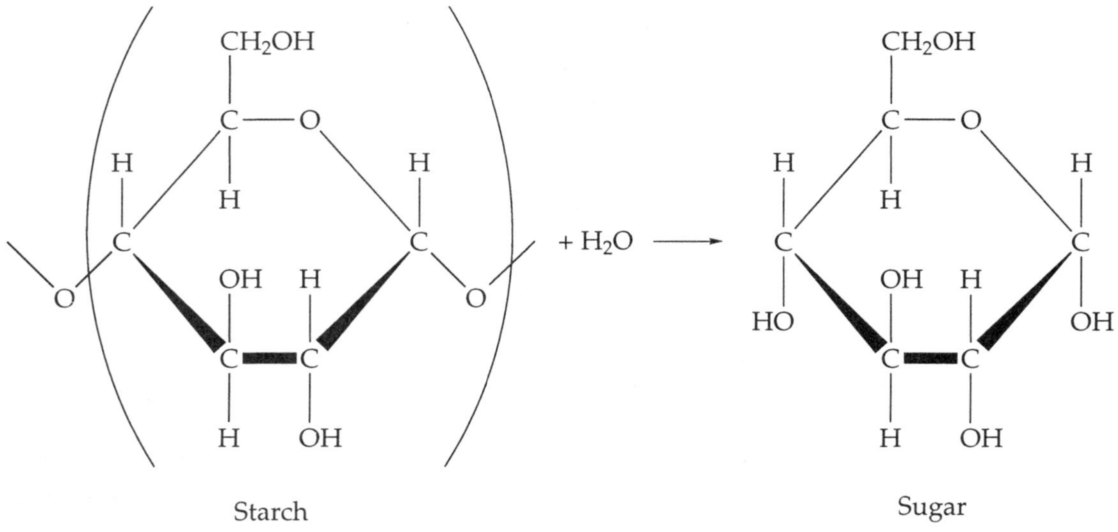

Figure 20.1
The hydrolysis of starch.

Table 20.1
Purpose of Solutions

Saliva Solution	Use in Experiment	Test Tube No.
A	Qualitative rate study	1–6
B	Effect of acid	7
C, D	Effect of heat	8

You will use four dilute saliva solutions, labeled A, B, C, and D, in this experiment. Their experimental purposes are outlined in Table 20.1. You can avoid confusion about the ultimate purpose of any solution by referring to this table.

Solution A will be used to examine the progress of the enzyme-catalyzed hydrolysis of starch during a period of 10 min. The activity of your salivary amylase will become apparent during this part of the experiment.

Solution B will be used to determine the effect of acidity on the active enzyme, and solutions C and D to determine the effect of heat. If acid or heat makes your amylase totally inactive, hydrolysis will no longer occur.

Experimental

Equipment: 9 small (13 mm × 100 mm) test tubes
5- or 10-mL graduated cylinder
50-mL Erlenmeyer flask
2 small (100-mL) beakers
250-mL beaker
heating plate
disposable rubber gloves
marking pencil

Materials: 6.0 M HCl
iodine solution
1% starch solution

Safety Precautions

A 6.0 M solution of hydrochloric acid can cause chemical burns in addition to ruining your clothes. If you spill any of this solution on you, wash the contaminated area thoroughly and report the incident to your laboratory instructor. You may require further treatment.

Because you will be handling body fluids (saliva) during this lab, special precautions should be taken: use only your own saliva solution during the experiment (or watch a partner if you are unable to use your own), wear disposable gloves when handling any glassware, and wash all glassware thoroughly when the experiment is completed.

Procedure

Getting Started

1. Obtain about 2 mL of your saliva in a 5- or 10-mL graduated cylinder.
2. Transfer the saliva to a clean Erlenmeyer flask containing 12 mL of distilled water. Agitate the flask thoroughly to get a homogeneous solution.
3. Mark 2 clean 100-mL beakers with identifying letters A and B.
4. Obtain 9 small test tubes. Mark 8 of them 1 through 8. Add 1.0 mL of distilled water to each, and mark the height of the water with a marking pencil. These marks will allow you to add 1.0 mL of solution quickly without using a graduated cylinder. Pour the water into a sink.
5. Mark the remaining test tube C.
6. Place a 250-mL beaker containing approximately 150 mL of distilled water on the heating plate. Heat the water to a gentle boil.

Preparing Solution C

1. Add 1.0 mL of the saliva solution to test tube C. Place the test tube in the beaker of boiling water.
2. Continue to heat the beaker for 20 min. During this time, go on to the next part of the experiment. When the 20 min has elapsed, remove the test tube from the water bath and allow the saliva solution to cool. This is solution C.

Working with Solution A

1. Obtain 6 mL of a 1% starch solution in beaker A. This is solution A.
2. Add 1 drop of the iodine solution and 1 drop of 6.0 M HCl to test tubes 1 through 6.
3. Pour 1.0 mL of solution A into test tube 1, using the mark that you have made. Save this test tube and its contents. You will use it as a reference throughout the experiment.
4. Read steps 5 through 9 before proceeding.
5. Add 1.0 mL of the saliva solution to solution A, using a 10-mL graduated cylinder and noting the time to the nearest second. Swirl the solution immediately and thoroughly.
6. When 1 min has elapsed, pour 1 mL of the solution from step 5 into test tube 2. Shake the test tube gently.
7. Repeat step 6 after 2, 4, 6, and 10 min, using test tubes 3, 4, 5, and 6, respectively.
8. Compare the colors in test tubes 1 through 6, and record the results. If the color has faded as time progressed, hydrolysis of starch has occurred. Record the degree of hydrolysis (none, some, or complete).
9. Save test tube 6 with its solution. You will use it as a reference throughout the remainder of this experiment. This test tube shows the maximum amount of hydrolysis that can occur during 10 min with your amylase.

Working with Solution B

1. Obtain 6 mL of the 1% starch solution in beaker B. This is solution B.
2. Add 4 drops of 6.0 M HCl to solution B.
3. Add 1.0 mL of the saliva solution to solution B. Note the time while swirling to obtain a homogeneous solution.
4. Set the solution from step 3 aside for exactly 10 min.
5. Add 1 drop of iodine solution to test tube 7.
6. After 10 min has elapsed, pour 1 mL of the solution from step 3 into test tube 7. Shake the test tube gently.
7. Record the color in test tube 7. Compare the color with those in test tubes 1 and 6. Has any hydrolysis occurred?
8. Wash, rinse, and dry beakers A and B. Identify one of them with the letter D.

Working with Solutions C and D

1. Obtain 6 mL of 1% starch solution in beaker D. This is solution D.
2. When solution C has cooled to room temperature, add it to solution D, noting the time. Swirl the solution.
3. Set solution D aside for exactly 10 min.
4. Add 1 drop of iodine solution to test tube 8.
5. After 10 min has elapsed, pour 1 mL of the solution from step 2 into test tube 8. Shake the test tube gently.
6. Record the color in test tube 8. Compare the color with those in test tubes 1 and 6. Has any hydrolysis occurred?

20
Amylase: One of Your Enzymes

NAME ____________________

SECTION ____________________

DATE ____________________

Prelaboratory Assignment

1. Define the following terms.
   a. enzyme

   b. catalyst

   c. amylase

   d. hydrolysis

2. How will you know whether the hydrolysis of starch has or has not occurred?

3. What are the purposes of the contents of the 8 test tubes used in this experiment?

4. What special safety precautions are cited in this experiment?

NAME _____

SECTION _____

DATE _____

Results

1. *Solution A*

Test Tube No.	Reaction Time (min)	Description of Color	Degree of Hydrolysis
1	0	_____	_____
2	1	_____	_____
3	2	_____	_____
4	4	_____	_____
5	6	_____	_____
6	10	_____	_____

2. *Solution B*

Test Tube No.	Reaction Time (min)	Description of Color	Degree of Hydrolysis
7	10	_____	_____

3. *Solutions C and D*

Test Tube No.	Reaction Time (min)	Description of Color	Degree of Hydrolysis
8	10	_____	_____

Questions

1. a. Summarize the effects of acidity and heat on your amylase.

   b. What reason can you give for the effect of acidity and heat on the activity of your amylase?

Appendix:
Significant Figures
and Rounding

Counting Significant Figures

Follow these rules when you need to count the number of significant figures in a measured quantity:

1. All digits are significant except zeros at the beginning of a number and possibly zeros at the end of a number. Thus, 5.46 cm, 0.546 cm, and 0.00546 cm all contain three significant figures.
2. Zeros at the end of a number are significant if the number contains a decimal point. Each of the following has four significant figures: 14.10 cm, 141.0 cm, and 1400. cm.
3. Internal zeros are significant. Thus, the zero in 10.3 is significant, and there are three significant figures in 10.3 cm.
4. End zeros may or may not be significant if the number has no decimal point. Ambiguity can be removed by using scientific notation. Or, if appropriate, add a decimal point.

Calculations Using Significant Figures

Measured quantities are often used in calculations. After we complete the calculation, how many significant figures should appear in the answer? In general, the **precision** (the dispersion or closeness of agreement between successive measurements of the same quantity) of an answer to a calculation cannot exceed the precision of the measured quantities used in the calculation. Apply two different and distinct rules to achieve this result:

1. When measured quantities are *multiplied* or *divided,* there should be as many significant figures in the answer as there are in the measurement with the fewest significant figures. For example:

$$115.36 \text{ cm} \times 3.52 \text{ cm} = 406 \text{ cm}^2$$

    In this calculation 3.52 has the fewest significant figures (three); therefore the answer should be reported to three significant figures (406 cm^2).
2. When the measured quantities are *added* or *subtracted,* a different rule applies. There should be the same number of decimal places in the answer as there are in the measurement with the smallest number of decimal places. For example:

$$103.1 \text{ cm} + 0.334 \text{ cm} = 103.4 \text{ cm}$$

    The quantity 103.1 cm has only one decimal place whereas 0.334 cm has three; therefore the answer should have only one decimal place, 103.4 cm.

It is important to note that any number whose value is known exactly will not affect the number of significant figures in a calculated result. For example, there are exactly 1×10^2 cm in 1 meter (m), so

$$5.243 \text{ m} \times \frac{1 \times 10^2 \text{ cm}}{1 \text{ m}} = 5.243 \times 10^2 \text{ cm}$$

The number of significant figures in the answer is determined by 5.243, not by 1×10^2.

Rounding

Rounding is often required to obtain the correct number of significant figures. Follow this general procedure.

1. If the digit to be dropped is less than 5, simply drop the digit. Some examples:

    12.44 rounded off to three significant figures gives 12.4

    1.2143 rounded off to three significant figures 1.21

2. If the digit to be dropped is 5 or greater, increase the preceding digit by one. Some examples:

    0.3568 rounded off to three significant figures gives 0.357

    1.2151 rounded off to three significant figures gives 1.22

In doing a calculation of two or more steps, it is typically best to wait and round off only the final answer. In other words, do not round off after each arithmetic operation. This ensures that small errors from rounding do not accumulate in the final number.

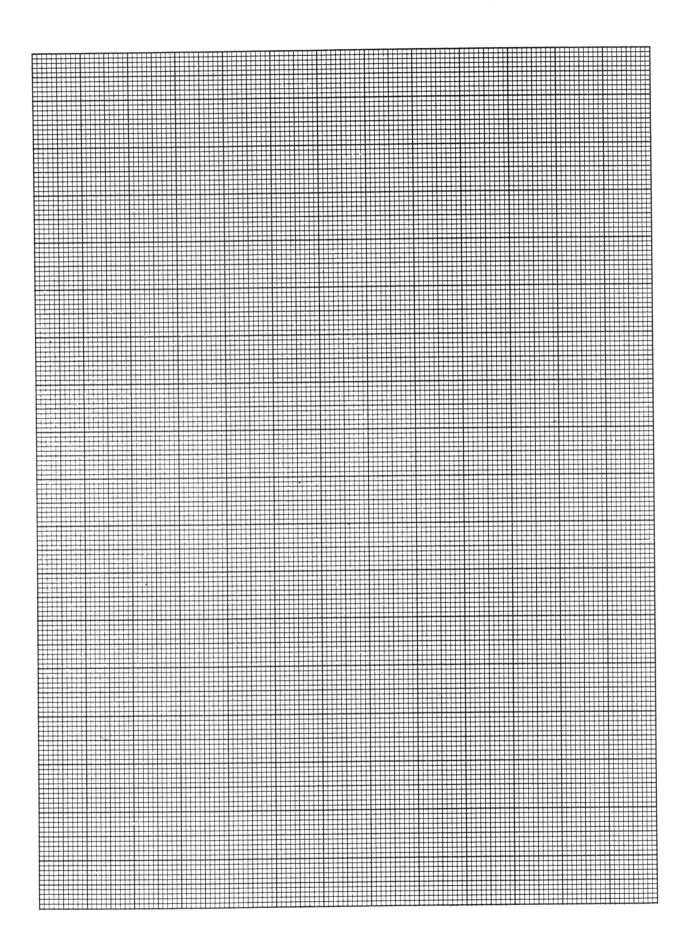

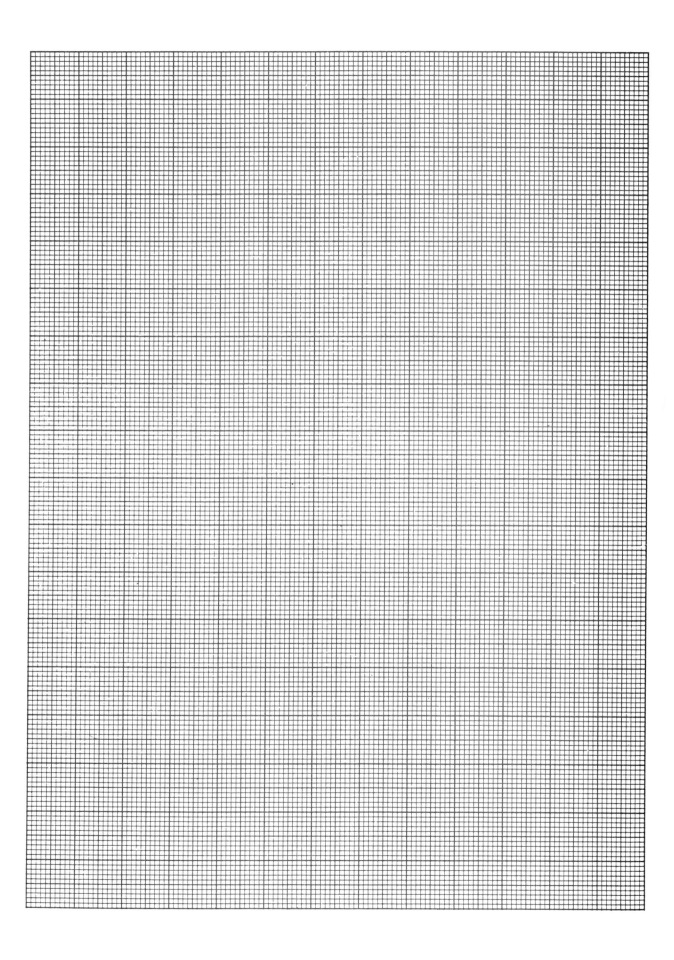

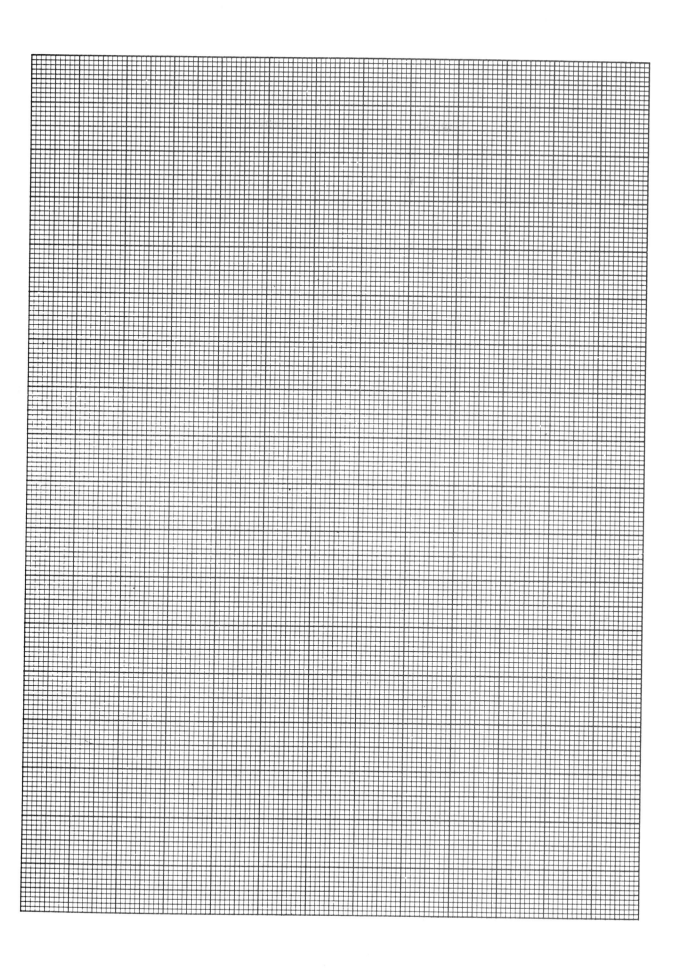

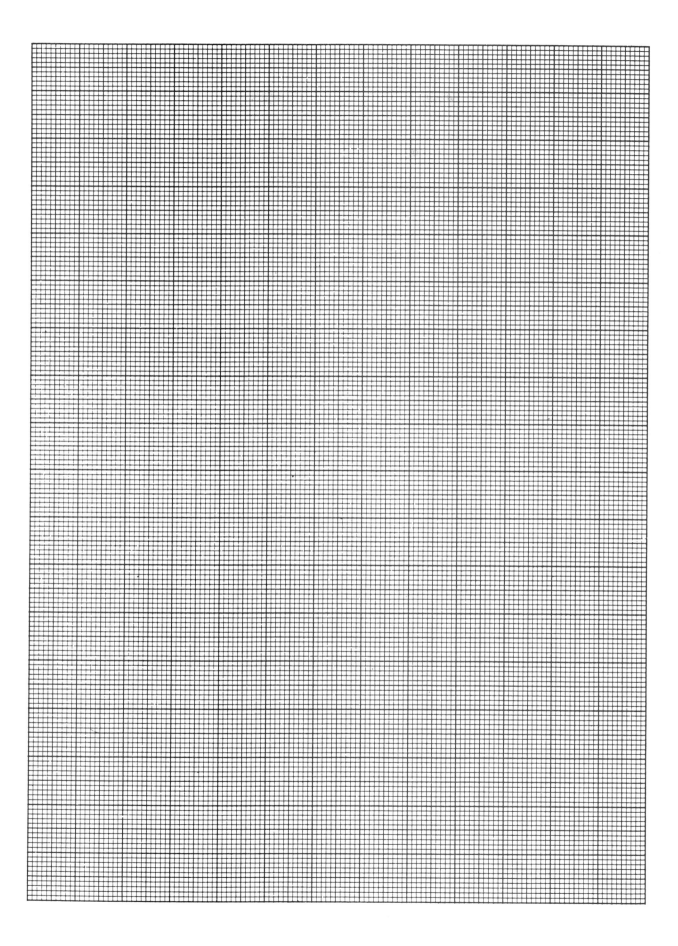

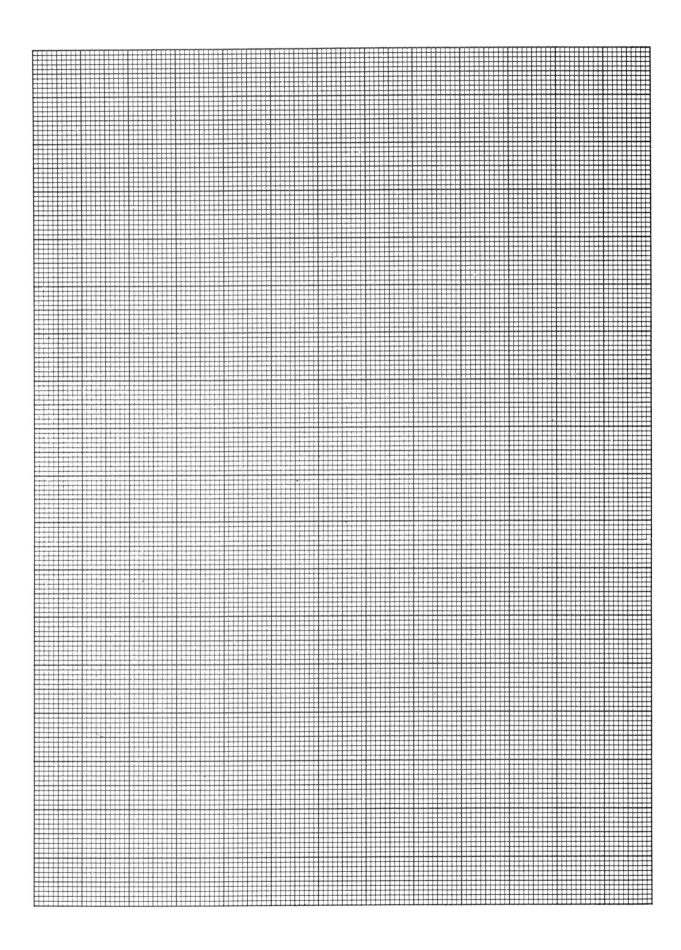

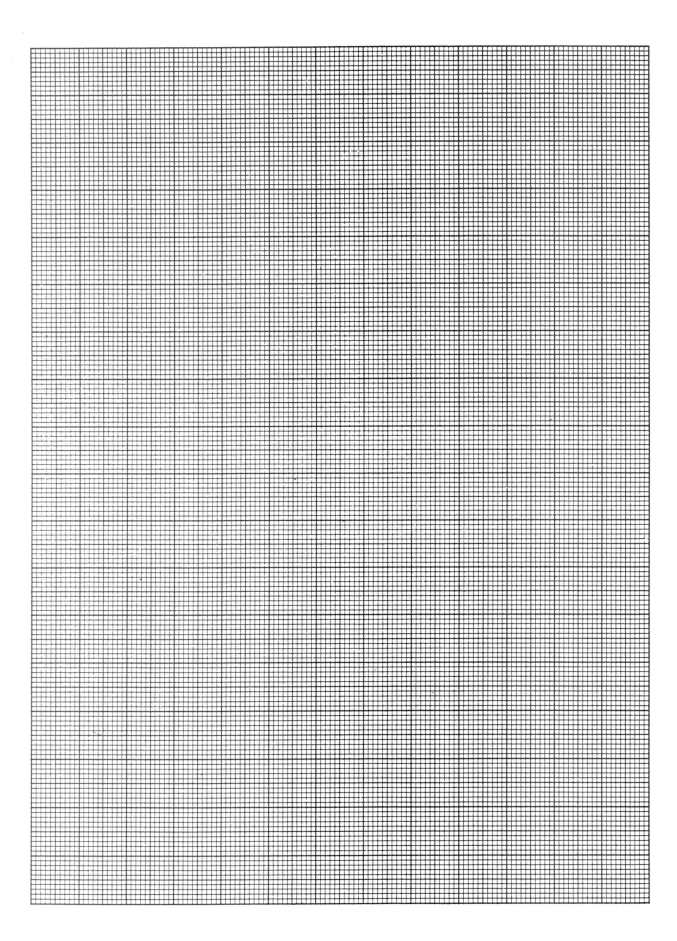